CITES and Cacti

a user's guide

Maurizio Sajeva[1], H. Noel McGough[2], Lucy Garrett[3], Jonas Lüthy[4], Maurice Tse-Laurence[2], Catherine Rutherford[2] and Giulia Sajeva[5]

1 Dipartimento di Biologia Ambientale e Biodiversità, University of Palermo, Italy
2 Royal Botanic Gardens, Kew, UK
3 School of International Development, University of East Anglia, Norwich, UK
4 CITES Management Authority of Switzerland, Bern, Switzerland*
5 Dipartimento Studi su Politica, Diritto e Società, University of Palermo, Italy

*address at time of preparation of guide

Kew Publishing
Royal Botanic Gardens, Kew

INTRODUCTION

The aim of '*CITES and Cacti: a user's guide*' is to provide a 'user friendly' tool for training those who wish to know about CITES and its application to the cactus family.

This guide has expanded speaker's notes for many of the PowerPoint slides included to provide a comprehensive coverage of particular topics. Most importantly, based on feedback from users of previous guides we have printed the guide in full colour and in larger format. We hope that this pack will allow users to tailor-make their own presentation on CITES and cacti and also form a concise but comprehensive reference. Please use this training tool and forward your comments to us so we can revise future editions to suit your needs.

Noel McGough,
Head of the Conventions and Policy Section,
UK CITES Scientific Authority for Plants,
Royal Botanic Gardens, Kew.

HOW TO USE THIS PACK

This pack consists of slides and speaker's notes for a presentation on the Convention on International Trade in Endangered Species (CITES). Divided into five separate topic areas (Introduction to Cacti, Appendix I Cacti, Appendix II Cacti, Non-CITES Cacti and Implementing CITES for Cacti), it can be used and adapted according to the background, interests and needs of your audience. The slides have been drafted in general terms with the hope that they will remain current, and therefore of use, for the foreseeable future.

Suggested speaker's notes accompany each slide. These notes are more specific than the slides and reflect information current as of July 2012. We have written these for speakers to use verbatim, providing a useful starting point from which you can tailor the slides, and accompanying speaker's notes, to reflect the specific needs of your audience, the length of the presentation and your own personal style. For example, you could illustrate some slides with examples from your own region or institution, or supplement the slides with extra images, such as cartoons, photographs, or newspaper cuttings. Such measures will undoubtedly increase the impact of an individual presentation. In addition, the slides can be printed onto transparency sheets for use with an overhead projector, or printed as handouts from the Microsoft PowerPoint® file on the CD-ROM, and given out to an audience for information.

CD-ROM

The CD-ROM contains the following files:

- 'CITESCacti.pptx', a Microsoft PowerPoint® presentation containing the slides and speaker's notes. You will need Microsoft PowerPoint 2007® (or a more recent version) to view and customise this file.
- 'CITESCacti.pdf', an Adobe Acrobat® presentation. You cannot modify this presentation but it can be viewed in 'full screen' mode using Adobe Reader®. You will need Adobe Reader® to view this file (can be downloaded from www.adobe.com).
- 'CITESCactiPack.pdf', a full copy of the text for the pack including the introduction and speaker's notes. This allows you to view the complete electronic document as well as print off part or all of the pack.

It also contains a series of training tools to support the implementation of CITES and the Convention on Biological Diversity.

CONTENTS

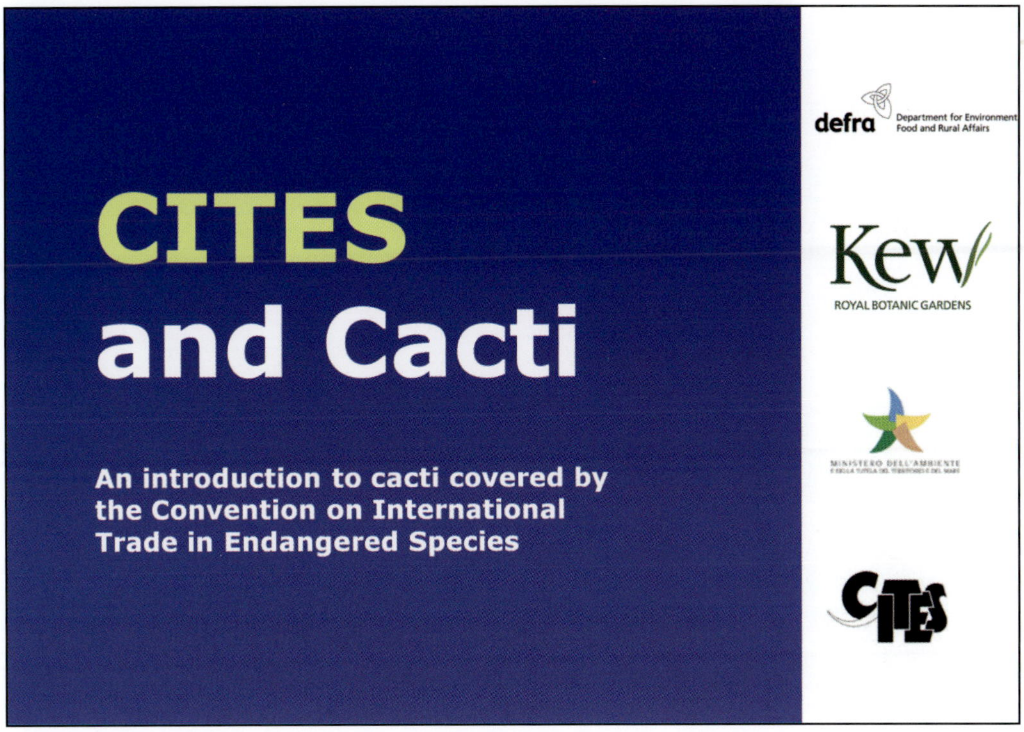

Slide 1: CITES and Cacti

The aim of this presentation is to introduce you to cacti covered by the Convention on International Trade in Endangered Species of wild fauna and flora – CITES – and to address some of the key issues concerning the implementation of the Convention for this important plant group. This is not an identification manual for cacti, but we will direct you to the most relevant and user-friendly work available. In all cases it is best to have contact with an expert on cacti to help you with identification issues. In this book we will outline the most important groups in trade, their level of propagation and the likely demand for wild plants in trade. We will also try to help you to distinguish between wild-collected plants and artificially propagated plants. In day-to-day CITES enforcement this is more important than identifying specimens to species level.

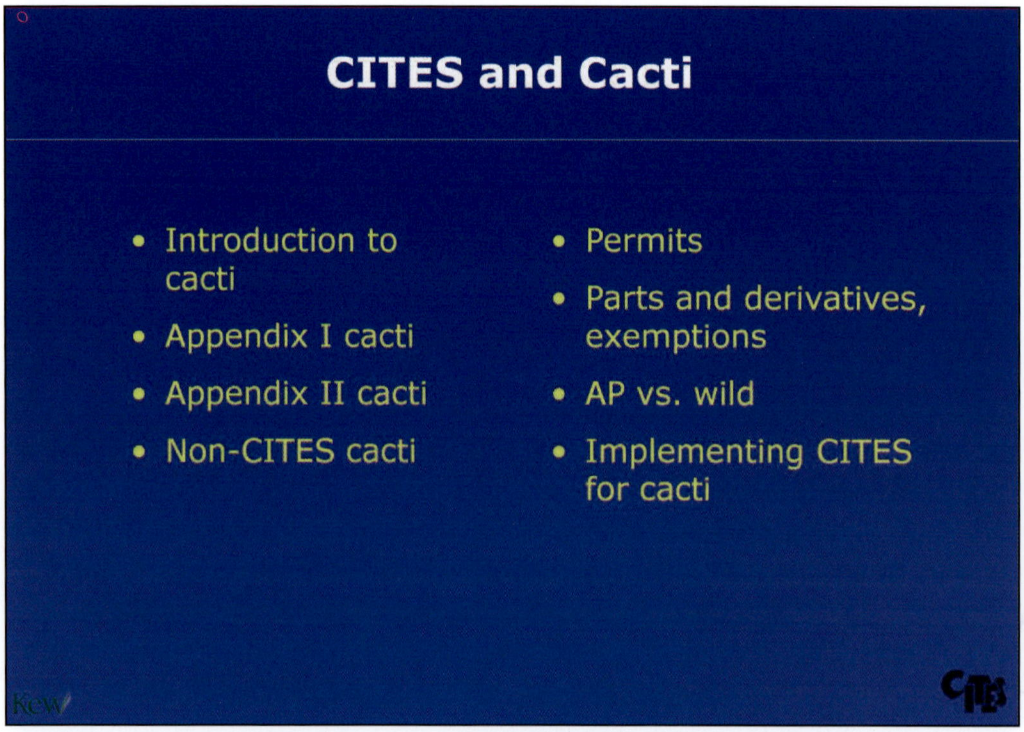

Slide 2: CITES and Cacti – What This Presentation Will Cover

This presentation will cover the following topics:

- Introduction to cacti
- Cacti on the CITES Appendices – Appendix I cacti
- Cacti on the CITES Appendices – Appendix II cacti
- Non-CITES cacti – leaf-bearing cacti, certain hybrids and cultivars
- Permits
- Parts and derivatives, exemptions
- Artificially propagated vs. wild
- Implementing CITES for cacti

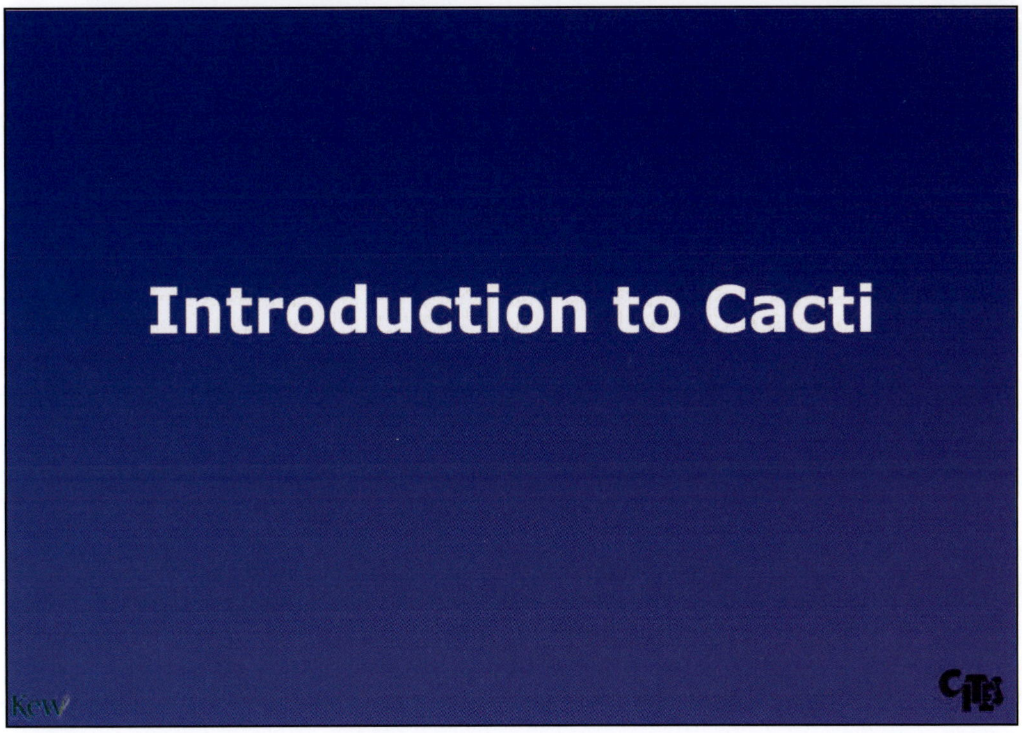
Introduction to Cacti

Slide 4: What are Cacti?

Cacti are a group of plants that botanists include in the cactus family or Cactaceae. They naturally occur, nearly exclusively, in the New World, from Canada to Patagonia, including the Caribbean. The Cactaceae family is a large and important family of succulent plants. Species range from the minute dwarf cacti hidden in the sands and gravel of the desert to the giant saguaro cacti, the well known backdrop to every cowboy movie and the target of cacti 'rustlers'. Virtually every North American and European home has had a cactus plant on its kitchen window sill – most likely a brightly flowered cultivar of *Schlumbergera,* the Christmas cactus.

[Note to speaker: the slide shows, top, from left: an epiphytic cactus, Rhipsalis ovata, cultivated in a hanging basket (App. II), a cactus nursery (Cactaceae spp., App. II), flower of Opuntia ursina, a prickly pear from the USA (App. II), Echinocactus grusonii, a top seller and propagated in millions, native to Mexico and classified as Endangered (App. II), Mammillaria formosa, native to Mexico and frequent in collections (App. II), cacti as an industrially produced commodity ('cactus blister') (Cactaceae spp., App. II), cacti as food in a Mexican supermarket, so called 'nopalitos', i.e. fresh shoots of Opuntia (App. II).]

Slide 5: Cacti Characteristics

Cacti are succulent plants and like all succulent plants possess tissue that is able to conserve water. This is normally a swollen stem and the stored water is used when there is little or no moisture available to the plant.

The vast majority of cacti do not have leaves. What botanists recognise as once being leaves have now been reduced to spines. The reduction of leaves into spines is an evolutionary adaptation to arid environments. The spines protect the plants from herbivores and are present in most species, but they have other functions as well. Spines may reflect light, shading the stem and reducing water loss or indeed trap water by condensing fog, as happens in the Atacama Desert in Chile. They may also act as camouflage as in *Sclerocactus papyracanthus.* In some cacti, the spines have been reduced or lost altogether and these cacti plants are small and hide in the soil or may produce chemicals that discourage animals from eating them. *Lophophora williamsii* (peyote) is a source of mescaline and can be used as an a hallucinogenic drug. In some countries the possession of this species is prohibited and it is illegal to extract the active substance, mescaline.

'Typical' cacti are characterised by globular or columnar and sometimes segmented stems which bear specialised felted discs called areoles from which spines develop. This feature is unique to cacti. A varying number of spines can arise from the areole and this number is often used as a characteristic to describe and distinguish different taxa of cacti in the field.

There are other succulent plants that can be confused with cacti, such as the succulent euphorbias. In *Euphorbia* the spines are usually in pairs, they never emerge from hairy tufts (areoles) and when damaged the stem produces a milky white latex which is an irritant to skin and eyes.

There are less familiar groups of cacti. Some are spiny shrubs which bear persistent leaves (pereskias) and others hang from rocks or branches of trees in humid climates, with tender, segmented, spineless stems ('epicacti', 'orchid cacti').

[Note to speaker: the slide shows, top left: Copiapoa cinerea (Chile). Below left: Matucana aureiflora (Peru). Top right: Euphorbia abyssinica (Africa). Below right: Euphorbia horrida (Africa)]

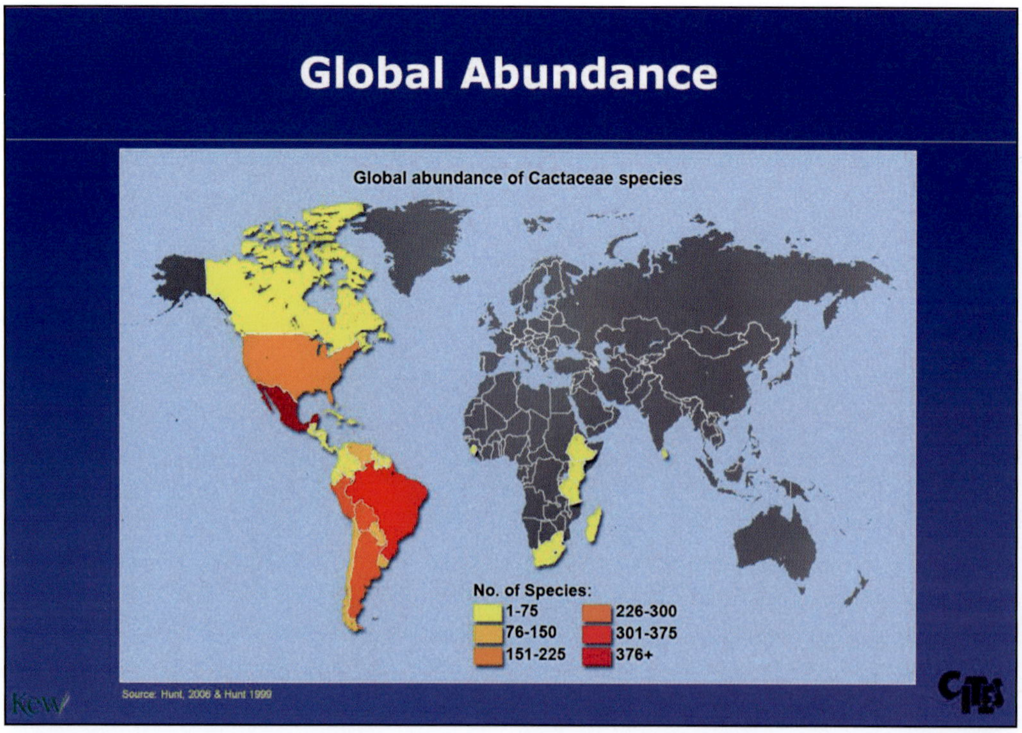

Slide 6: Global Abundance

Cacti are endemic to the Americas with the exception of just one genus, *Rhipsalis*, whose distribution stretches from South America to southern Africa and Sri Lanka. There are three 'hot spots' for species diversity. Prime amongst these is Mexico and the adjacent south-western USA where nearly 30% of cacti genera are endemic and nearly 600 species are native. The arid areas of the south-western Andes provide another hot spot, covering parts of Peru, Bolivia, Chile and Argentina. Eastern Brazil, including dry caatinga and the rocky high terrain of the campo rupestre, make up the final hot spot.

Some cacti, the 'prickly pears' (*Opuntia*), have been introduced to several regions of the world, such as Australia, southern Africa or the Mediterranean, mainly for their edible fruits and have since become invasive in many areas.

This slide illustrates the global abundance of *Cactaceae* species. Mexico has the highest abundance, followed by Brazil, Peru, Bolivia and Argentina.

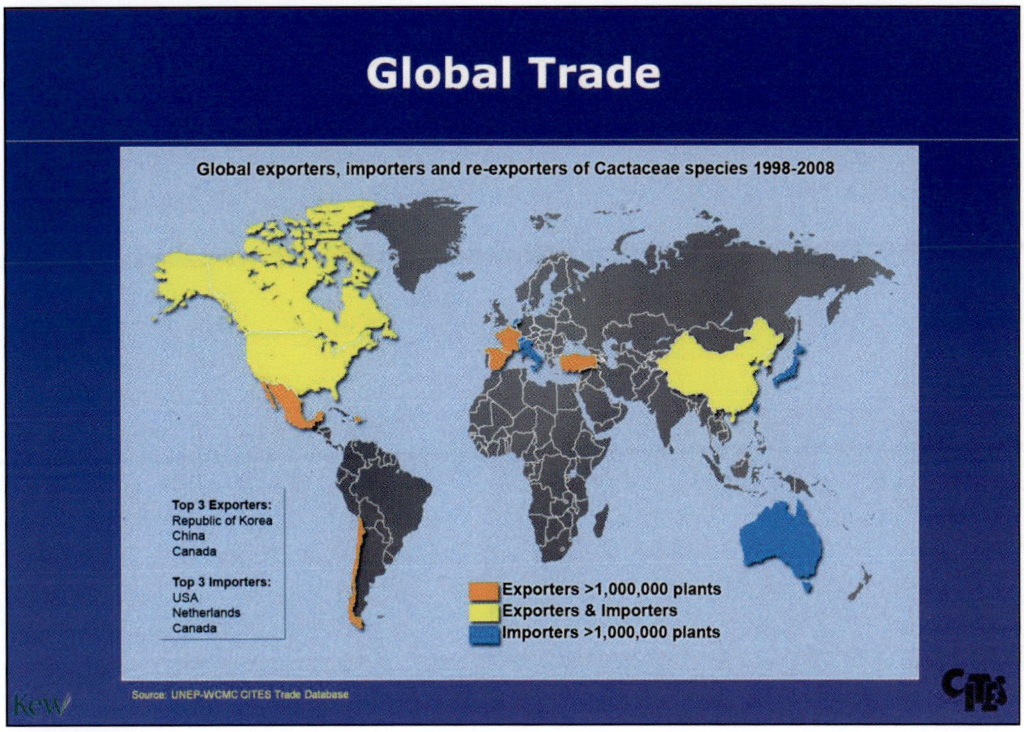

Slide 7: Global Trade

The first reports of cultivation of cacti in Europe date back to the 1500s, following their introduction from the Americas. Today North America, Europe and Asia produce many millions of propagated cacti per year from their horticultural industry. However, there remains a persistent demand for species collected from the wild.

Habitat destruction and illegal collection for international trade continues to endanger cacti, with new threats emerging all the time. For example, the demand for desert plants for use in landscaping has fuelled the consumer market in succulent plants. TRAFFIC North America estimates that between 1998 and June 2001 nearly 100,000 succulents, with an estimated value of US$3 million, were harvested from Texas and Mexico to supply the landscape garden market in Phoenix and Tucson.

Newly discovered and described species continue to appear in illegal trade worldwide because of high prices and demand by collectors.

Using CITES trade data, this slide illustrates major exporters, importers and exporters-importers of live, artificially propagated and wild Cactaceae species between 1998-2008 (countries trading in quantities of more than one million plants are shown). The Republic of Korea, China and Canada are the top three exporters, with the United States of America, the Netherlands and Canada the top three importers. Trade quantity is measured by number of live plants, seeds and stems. This figure does not include data of trade in other parts and derivatives (e.g. carvings, fruit, extract, powder, roots etc) nor trade recorded in other units (e.g. kg, g, m^3, m^2, shipments, sets etc.).

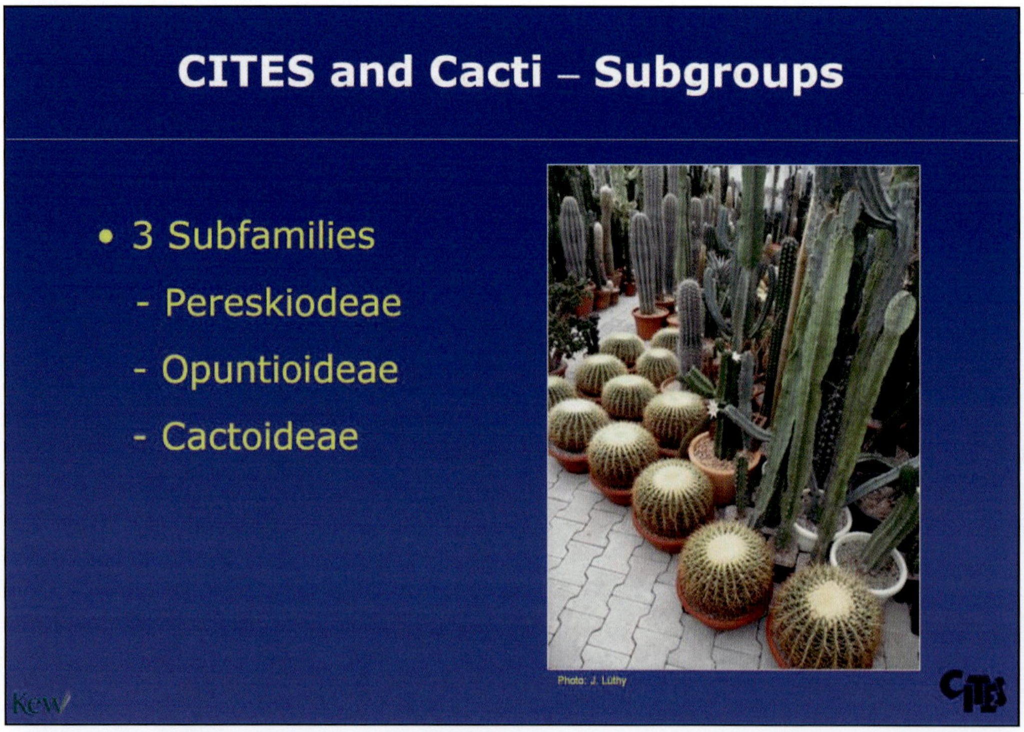

Slide 8: CITES and Cacti – Subgroups

The cactus family is included in CITES Appendix II, with a number of the most endangered species listed in Appendix I. There are a small number of exemptions to the full family listing where wild and artificially propagated plants of three genera are excluded from CITES control, due to the fact that they are not threatened by international trade and that they are easily distinguished from other cacti as they are leaf-bearing shrubs and trees.

The Cactaceae family is usually divided into three subfamilies:

1. Pereskiodeae includes the cacti with the most primitive or ancestral features. Members of this subfamily have leaves and are quite different from the species involved in international trade. The genus *Pereskia* is excluded from CITES control.

2. Opuntioideae includes species more or less similar to the prickly pear. There is little trade within this subfamily. Within this group the leaf-bearing genera *Pereskiopsis* and *Quiabentia* are not CITES controlled.

3. Cactoideae is the subfamily with the most diverse habits, shapes and dimensions. All the traded species are included in this subfamily and all genera are CITES regulated. In addition to the very popular globular and columnar cacti, the sometimes neglected epiphytic cacti belong to this group.

[Note to speaker: the slide shows Echinocactus grusonii (globular), Cereus spp. (columnar to the right) and Carnegiea spp.]

Slide 9: Appendix I Cacti – What is Controlled?

The commercial trade in specimens of Appendix I plants taken from the wild is effectively prohibited. Trade in artificially propagated specimens is allowed, but is subject to the granting of permits.

For CITES Appendix I, in addition to the plants themselves, the trade (international movement) in any part of the plant or any readily recognisable product made from them is also subject to CITES control. This includes seeds and scientific material such as herbarium specimens, material preserved in spirit, such as flowers or fruit, and DNA samples.

Slide 10: Appendix II Cacti – What is Controlled?

The complete cactus family is included in CITES Appendix II, with a number of the most endangered species listed in Appendix I. There are a small number of exemptions to the full family listing where wild and artificially propagated plants of three genera are excluded from CITES control, due to the fact that they are not threatened by international trade and that they are easily distinguished from other cacti as they are leaf-bearing plants. These genera are *Pereskia*, *Pereskiopsis* and *Quiabentia.*

The family listing of Cactaceae is also annotated in the CITES Appendices. These annotations indicate what parts and derivatives are controlled by CITES. The Appendix II listing of cacti is currently annotated as follows by the numbers 6 and #4. The annotation 6 indicates that artificially propagated hybrids and/or cultivars of *Hatiora* x *graeseri*, *Schlumberga* x *buckleyi*, *Schlumberga russelliana* x *Schlumberga truncata*, *Schlumberga orssichiana* x *Schlumberga truncata*, *Schlumberga opuntioides* x *Schlumberga truncata, Schlumberga truncata* (cultivars), and Cactaceae species colour mutants grafted on the following grafting stocks: *Harrisia* 'Jusbertii', *Hylocereus trigonus* or *Hylocereus undulatus*, and *Opuntia microdasys* (cultivars) are not controlled by CITES. The symbol #4 indicates that all parts and derivatives of Appendix II cacti are controlled except:

a) seeds, except those from Cactaceae species exported from Mexico
b) seedling or tissue culture obtained in vitro, in solid or liquid media, transported in sterile containers
c) cut flowers of artificially propagated plants
d) fruits and parts and derivatives thereof of naturalised or artificially propagated plants of the family Cactaceae
e) stems, flowers, and parts and derivatives thereof of naturalised or artificially propagated plants of the genera *Opuniia* subgenus *Opuntia* and *Selenicereus.*

All other material is CITES controlled. This includes leaves and parts of leaves, Herbarium specimens, DNA, and powdered stem for use in industry or pharmacy. There is one additional exemption concerning rainsticks and this will be dealt with elsewhere.

Appendix I Cacti

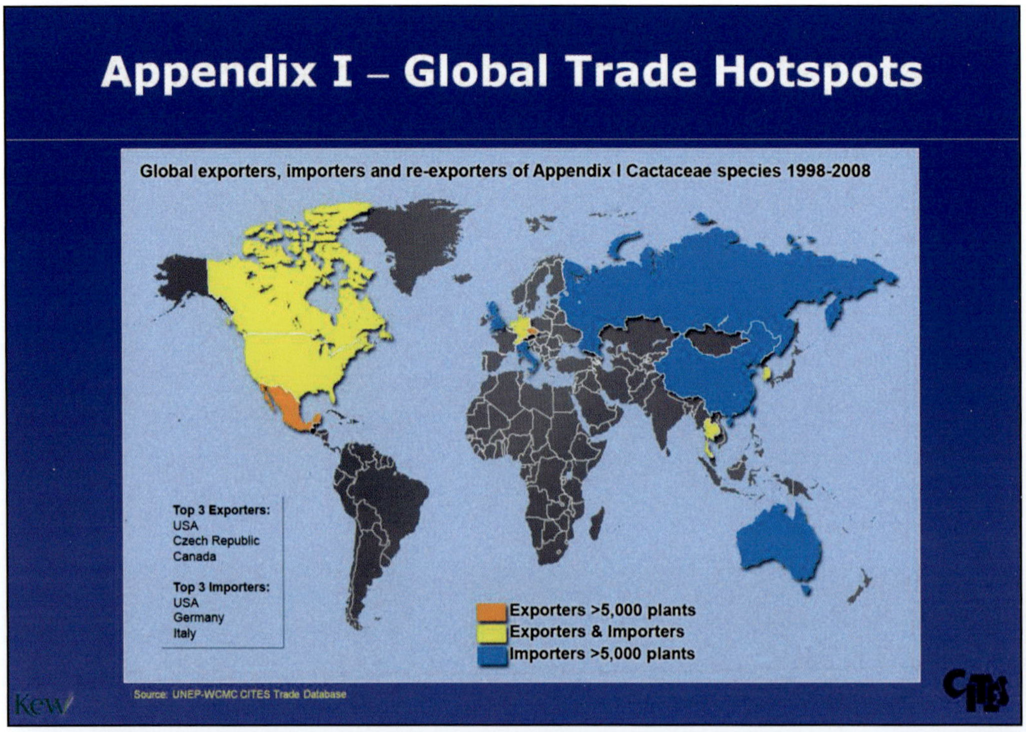

Slide 12: Appendix I – Global Trade Hotspots

Using CITES trade data this slide illustrates global exporters and importers of live, artificially propagated Appendix I cacti species between 1998-2008 (countries showing trade in more than 5,000 plants). Countries exporting cacti are illustrated in orange, those importing are illustrated in blue and countries that export and import cacti are illustrated in yellow. The top three exporters are the United States of America, the Czech Republic and Canada. The top three importers are the United States, Germany and Italy.

Trade quantity is measured by number of live plants, seeds and stems. This figure does not include data of trade in other parts and derivatives (e.g. carvings, fruit, extract, powder, roots etc) nor trade recorded in other units (e.g. kg, g, m^3, m^2, shipments, sets etc).

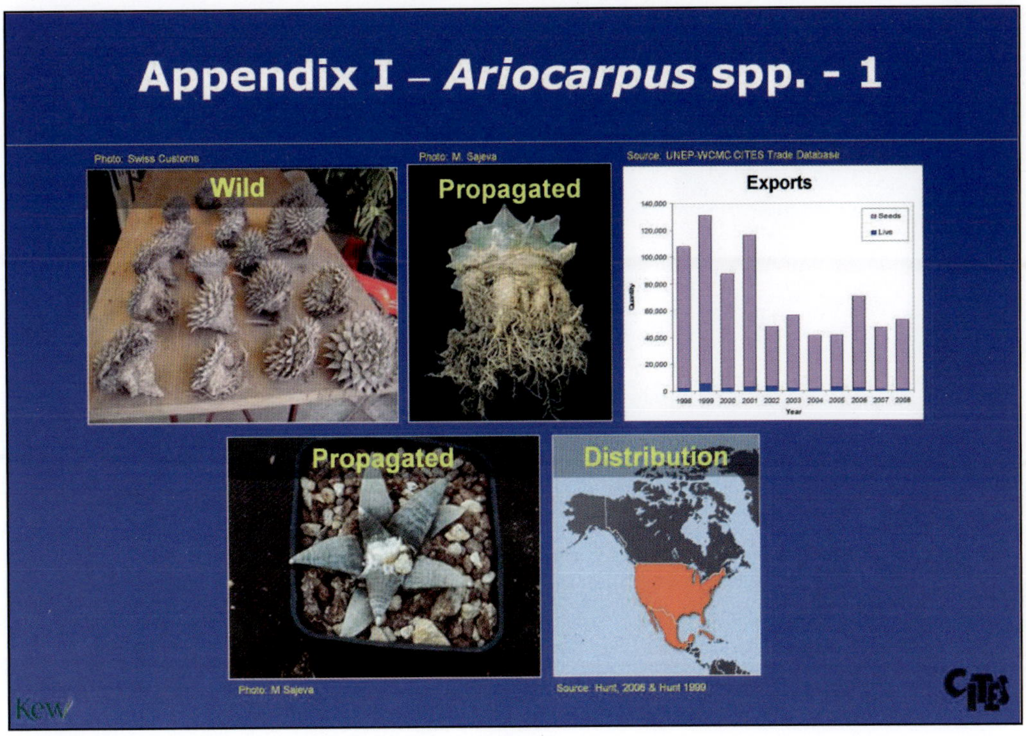

Slide 13: Appendix I – *Ariocarpus* spp. - 1

The genus *Ariocarpus* includes seven species and one subspecies and is virtually confined to the desert areas of the north Mexican highlands. Six taxa are considered endangered. One species, *Ariocarpus fissuratus,* not considered threatened, also occurs in southern Texas, USA. Plants mostly grow underground with stout roots and tubercles (wart-like protuberances) arising from the soil. Spines are usually absent, except in seedlings. The plants resemble living rocks when seen in their natural habitat.

All species of *Ariocarpus* are included in Appendix I. They are slow-growing, and it takes several years for them to reach flowering size and many more years to reach full specimen size. Seeds are available and propagation is easy (but slow and thus expensive). There is high demand for large specimens and illegal collection in habitat causes a severe decline of the wild populations.

Illegal trade still occurs, sometimes at a highly organised commercial scale, but most frequently at a low level, mainly confined to specialist collection by visiting collectors exported in luggage or by mail or courier.

CITES trade data is virtually confined to seed, with a few artificially propagated live plants. A small amount of wild exports were reported in 2000. *Ariocarpus retusus* is most frequently seen in trade. Switzerland, the United States of America and the Czech Republic are the main exporters.

[Note to speaker: this slide shows, top left: (illegally) wild-collected Ariocarpus retusus. Note that roots are truncated. Top right: artificially propagated seedling of Ariocarpus retusus with bare roots. Note that the roots are intact. Bottom left: Ariocarpus retusus, artificially propagated juvenile plant. Note that the tubercles of juvenile plant are narrower than in adult plants and the colour is greenish, compared with greyish-brownish of wild-collected plants.]

Slide 14: Appendix I – *Ariocarpus* spp. - 2

Ariocarpus bravoanus subsp. *bravoanus* is a rare local endemic which was first described in 1992. First known from only one site in Mexico, after its description there was significant illegal trade which greatly diminished the population at its original site. No legal exports of commercial material from Mexico have been reported. Mature adult material in trade is likely to be wild in origin. Wild-collected specimens may be found in private collections. It has a conservation rating of Critically Endangered.

Ariocarpus bravoanus subsp. *hintonii* (Vulnerable) is another local endemic and has been widely collected from its habitat since it was first recorded in 1981, substantially depleting wild populations. Significant illegal trade has occurred with no known exports of legal material for the international market. Artificially propagated seedlings and young plants are now in trade but large old plants are likely to be wild in origin.

Ariocarpus fissuratus is distributed in Mexico and the United States of America. It has been illegally collected and traded in the USA, advertised on the internet and also exported to some extent, e.g. in post/courier parcels with false declarations.

Ariocarpus retusus (Least Concern) has been smuggled from Mexico to the United States of America, from where it has been exported in post/courier parcels with false declarations. It is also in trade as a range of subspecies based on wild material collected at the population level and displaying a range of variation. These subspecies are of doubtful taxonomic merit. This material is in trade as novelties, attractive to collectors and their collection is heavily impacting the local populations. Names used include *A. confusus*, *A. retusus* subsp. *scapharostroides*, subsp. *jarmilae*, subsp. *horaceki*, subsp. *panarottoi.*

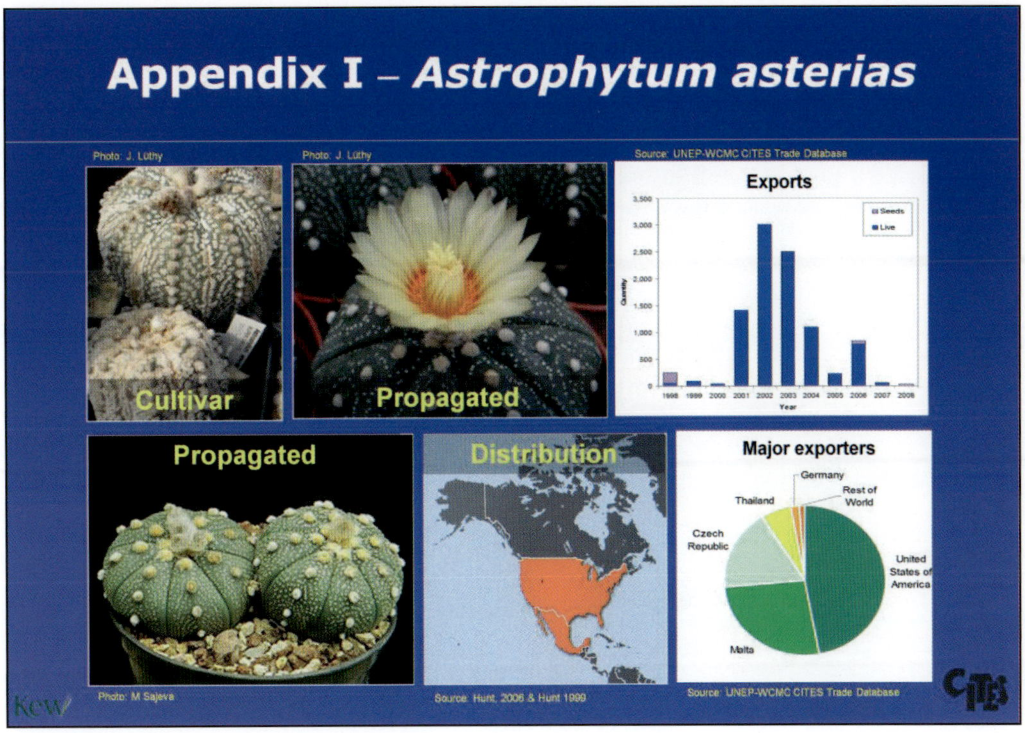

Slide 15: Appendix I – ***Astrophytum asterias***

This is the only species of the six in the genus *Astrophytum* listed on Appendix I. *Astrophytum asterias* is found in Mexico (Nuevo León, Tamaulipas) and the USA (Texas) and is has a conservation assessment of Vulnerable. Plants are completely spineless, flattened globes, with woolly tufts (areoles) arranged in lines down the ribs and covered with tiny white wool flecks. They are usually deeply embedded in the ground in their habitat. Their shape and sunken habit gives rise to their common name of star cactus. They are quite easy to grow from seed and are relatively fast-growing.

There has been large scale illegal collection in the past, but now large numbers of artificially propagated specimens are available in the marketplace. There is little or no demand for wild plants and illegal trade is unlikely unless a wild plant is particularly sought after. Increasingly, artificially propagated plants are sourced from Thailand and China, especially certain cultivars that strongly differ from the wild form (but still are Appendix I specimens).

In Mexico, *Astrophytum asterias* is often confused by local residents with *Lophophora williamsii*, the peyote cactus, and has been allegedly eradicated in some places for this reason. Many habitats are degraded because of land use for cattle breeding.

Especially in Japan, some cultivars such as 'Super Kaputo' are very popular. One individual is reported to have paid $35,000 for the first plant of the form 'Mirakuru'. They differ in having much more pronounced wool flecks. Other cultivars are variegated, showing yellow stem parts. Hybrids with other species of *Astrophytum* can be found in trade. They usually show spines and more acute ribs.

CITES trade data is confined to artificially propagated live plants and seeds, with no wild trade reported. The United States of America, Malta and the Czech Republic are the main exporters.

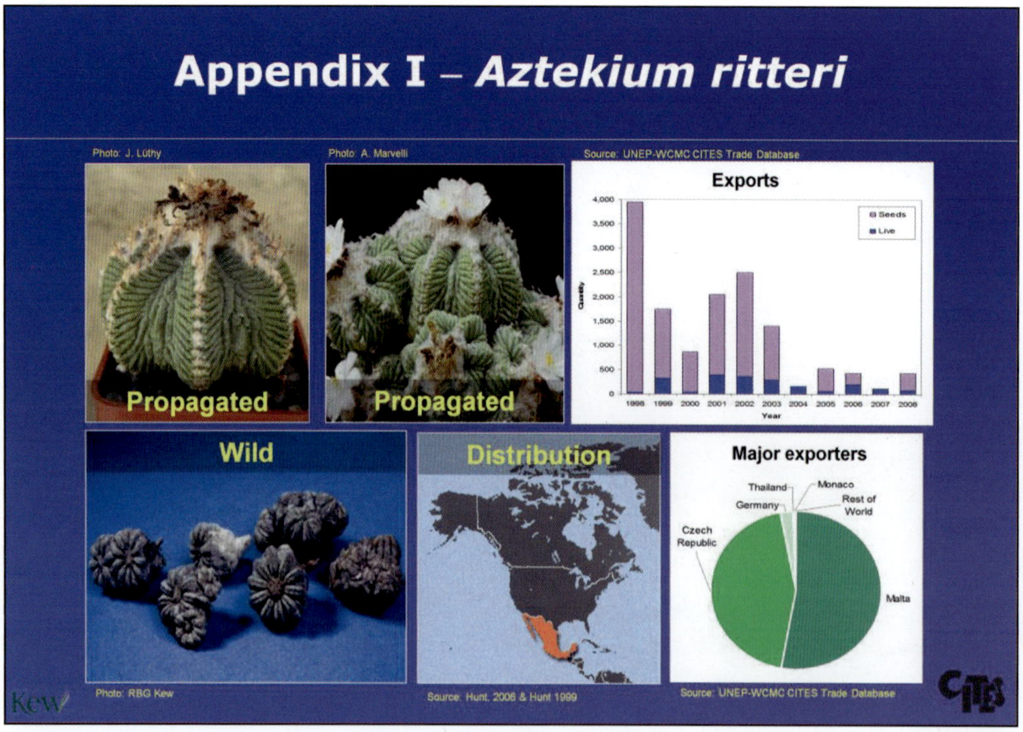

Slide 16: Appendix I – ***Aztekium ritteri***

The genus *Aztekium* includes two species, one of which, *Aztekium ritteri,* is included in Appendix I and has a conservation rating of Least Concern. It is confined to a small area in Nuevo León, Mexico. It has a simple or clustered grey-green, slightly flattened globular body which is ridged or ribbed with a coarse wrinkled appearance. The plant is nearly spineless.

It is one of the slowest growing cacti, annual growth being 1-2mm. Young seedlings are less than 1mm in diameter and are very difficult to keep alive. This explains very high prices even for small plants and creates a strong incentive for trade in wild-collected plants. It is a sought-after species, and in the past a large amount of wild-collected material has been illegally exported from Mexico to supply the international demand for this species. Seeds are available, but growth from seed to flowering plants takes several years. Grafting may speed up the growth process.

Illegal trade in wild-collected specimens still occurs, but at a low level, mostly confined to specialist collection by visiting collectors who export the plants in their luggage or by mail or courier.

The other species in the genus, *Aztekium hintonii*, caused a great stir in the cactus world when it was discovered with *Geohintonia mexicana*, a new genus of cactus growing on gypsum-rich deposits, in Nuevo León in 1991. Although these species were automatically included in the family listing of cacti on CITES Appendix II and Mexico bans exports of all wild cacti, including seeds originating from Mexico, considerable illegal trade followed. Illegal wild plants of these taxa are still in trade and can be found in many collections.

CITES trade data is virtually confined to artificially propagated live plants and seeds with no recent wild trade reported. Malta and the Czech Republic are the main exporters.

[Note to speaker: this slide shows, top left: grafted seedling. Top middle: old, grafted plant; grafting stimulates branching. Below: wild-collected plants; they are sparsely branching with age.]

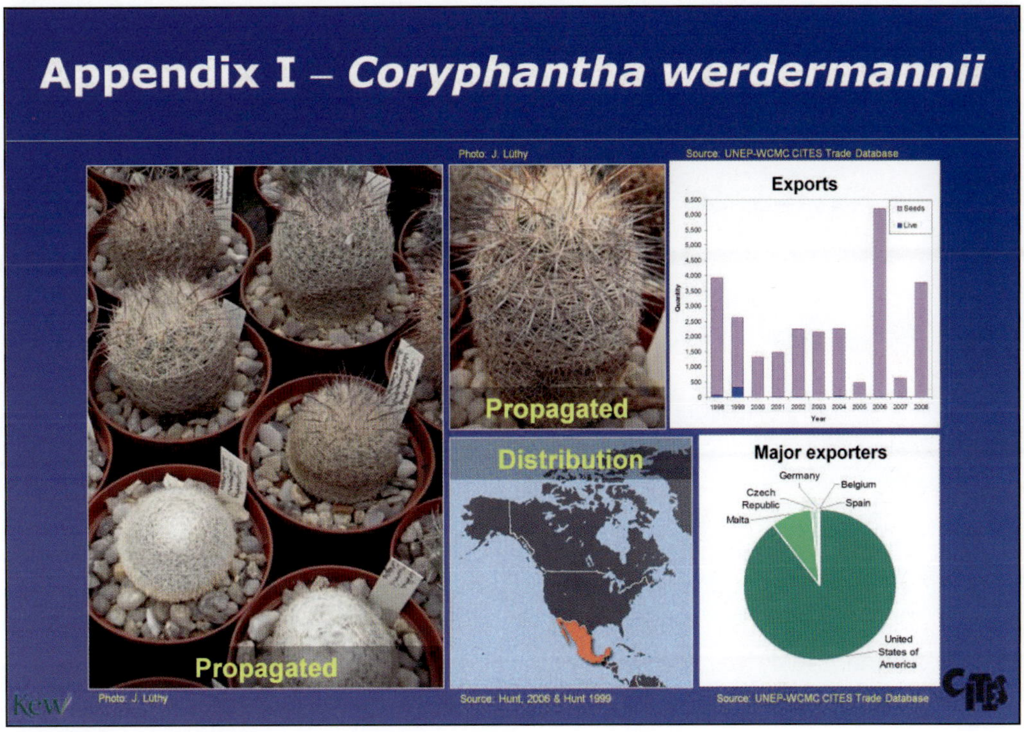

Slide 17: Appendix I – *Coryphantha werdermannii*

One of the 42 species of the genus *Coryphantha*, *C. werdermannii* occurs on limestone hills in Coahuila, Sierra de la Paila, Mexico. Plants are spherical but with age they become columnar. They grow to 10cm or more in height and up to 8cm in width. Plants are usually solitary. The outstanding feature of this species is an abrupt and contrasting change from the juvenile aspect with dense, flattened, white spines and a smooth appearance to the adult with additional, dark, protruding spines and a 'fierce' appearance. This habit attracts the attention of some collectors.

It is not often seen in trade and is of limited interest to collectors; illegal trade is not now reported. Export has occurred in the past, over a limited period of time and in limited numbers. The natural population is abundant and healthy and this species may be a suitable candidate for down-listing to Appendix II.

CITES trade data is virtually confined to a small amount of artificially propagated live plants and seeds. The United States of America and Malta are the main exporters.

[Note to speaker: this slide shows, left: juvenile (foreground) and change to adult spination. Juvenile and adult plants look like two different species, due to strong dimorphism.]

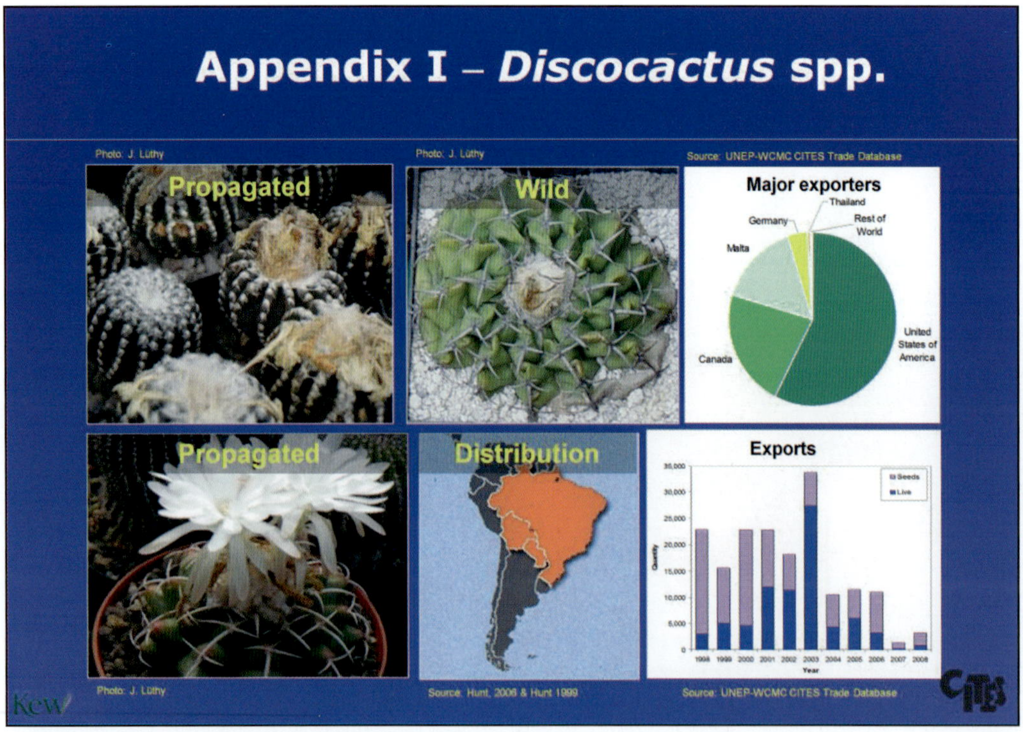

Slide 18: Appendix I – *Discocactus* spp.

The genus *Discocactus* includes 11 species, 7 of which have conservation ratings from Vulnerable to Endangered. They are located in the drier parts of Bolivia, Brazil and Paraguay. Plants are spherical and flowers arise from a distinctive hat-like, bristly flowering zone at the top of the plant (cephalium), a character in common with the genus *Melocactus*. The species of *Discocactus* cannot easily be distinguished from each other, with the exception of *Discocactus horstii* and therefore the whole genus has been included in CITES Appendix I.

The genus is not a major attraction for collectors, apart from the dwarf *Discocactus horstii* which is now widely available as propagated plants. There is some interest in collectors for cephalic cacti in general and Brazilian cacti in particular. There was a peak in demand in the 1970s and 1980s when a range of new species was described. The requirement of higher greenhouse temperatures in winter limits the interest of collectors. In addition, the flowers, although spectacular and scented, are ephemeral and nocturnal and fade the following morning.

Some illegal trade still occurs and care should be taken when examining imports from South America. Propagation by seed is quite easy but it takes time for the plants to produce the mature cephalium which is the major attraction to collectors. Cultivation of *Discocactus horstii* on its own roots is very difficult as the taproots are very sensitive to excessive moisture. For this reason the species is normally grafted.

Trade in *Discocactus* spp. is dominated by *D. zehntneri* and *D. zehntneri* subsp. *boomianus* and is in artificially propagated live plants and seeds. Trade has declined over the years 2006-2010.

[Note to speaker: this slide shows, top left: Discocactus horstii propagated by seed. Top middle: Discocactus placentiformis. Bottom: Discocactus spp.]

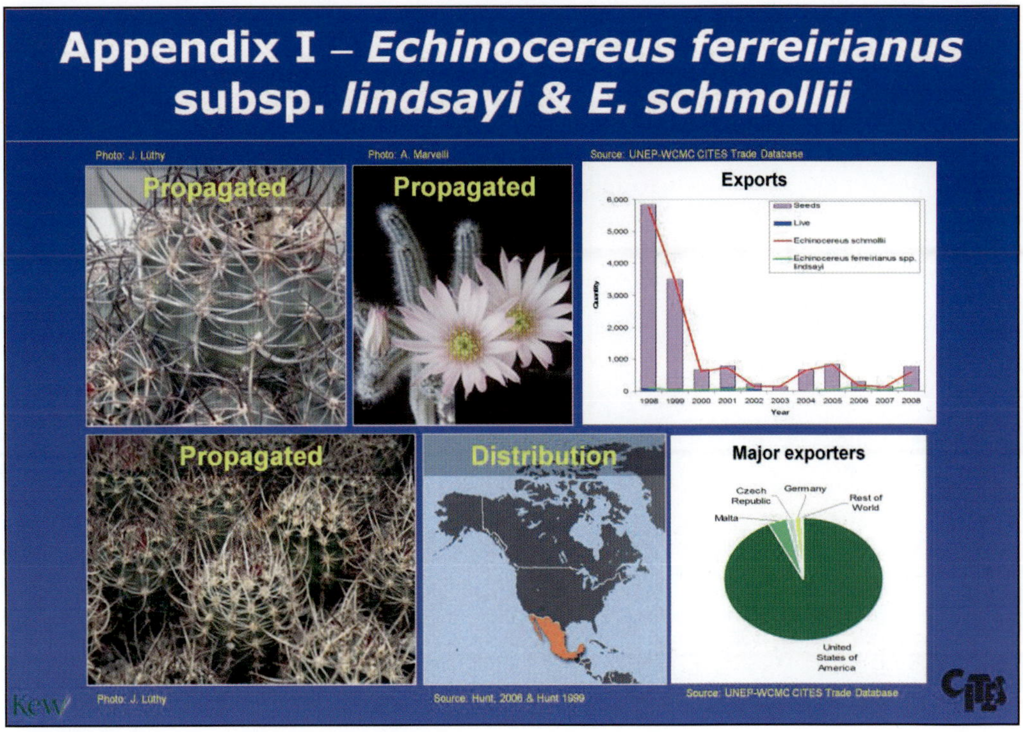

Slide 19: Appendix I – *Echinocereus ferreirianus* subsp. *lindsayi* & *E. schmollii*

The genus *Echinocereus* includes some 70 species. Only *Echinocereus ferreirianus* subsp. *lindsayi* and *E. schmollii* are included in CITES Appendix I.

Echinocereus ferreirianus subsp. *lindsayi* is a single-stemmed plant up to 13cm tall and 10cm in diameter, with tuberculate ribs and coarse prominent spines up to 10cm long. Known only from two sites in Baja California, Mexico, it was nearly wiped out at its original location. Today additional, undisturbed locations are known, but the subspecies is still rare with a conservation rating of Critically Endangered. In contrast, subspecies *ferreirianus,* (the typical subspecies), is quite widespread and has a conservation rating of Least Concern. It has more slender, clustering stems and more delicate, straight spines and is listed in Appendix II. This is an example of a so called 'split listing' within a species. Both taxa are now available as artificially propagated plants and illegal trade is unlikely.

Echinocereus schmollii has stems up to 25cm tall and 1-2cm in diameter. Called the lambs tail cactus due to its characteristic growth form, this thin columnar cactus normally grows intertwined with shrubs. Soft in consistency, the spines grow needle-like in older plants. It has a conservation rating of Critically Endangered. It is not frequent in international trade and the limited market is supplied by propagated plants. It is not likely to be in illegal trade.

Reported exports shown in this slide are the sum of both *Echinocereus ferreirianus* subsp. *lindsayi* and *E. schmollii*. Trade, predominantly in *E. schmollii,* is reported as artificially propagated live plants and seeds. Of this, most of the trade in *E. schmollii* is in seeds (97%) with only 37% trade in seeds for *Echinocereus ferreirianus* subsp. *lindsayi.*

[Note to speaker: this slide shows, top left: Echinocereus ferreirianus subsp. lindsayi. Bottom left: E. ferreirianus subsp. lindsayi. Top right: E. schmollii]

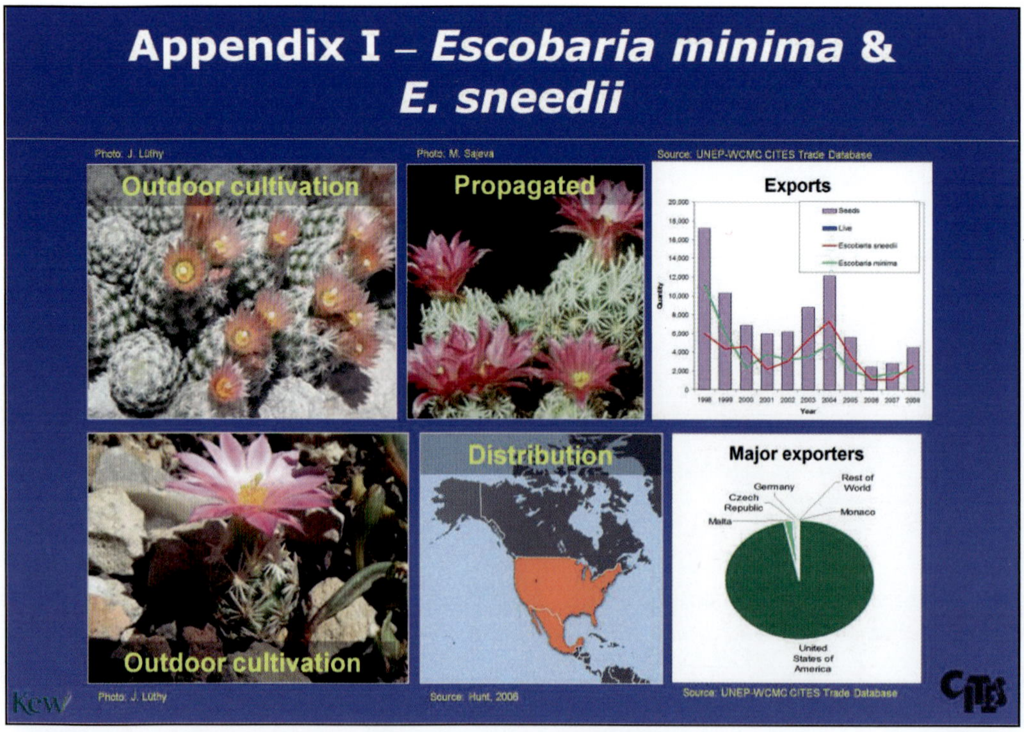

Slide 20: Appendix I – *Escobaria minima & E. sneedii*

The genus *Escobaria* includes 19 species distributed in Canada, Cuba, Mexico and the USA. Only two species are included in CITES Appendix I: *Escobaria minima* and *Escobaria sneedii* (subsp. *sneedii* and subsp. *leei*). They are usually included in the genus *Coryphantha* in American literature and trade.

Escobaria minima is an endemic of Brewster County in Texas, USA. It is a distinctive dwarf, spherical cactus 3-4cm tall and 1-2cm wide. Originally sought after for its dwarf habit and rarity, it is now widely held in collections and available as propagated, often grafted, plants. It has a conservation rating of Vulnerable. It is unlikely to be in illegal trade.

Escobaria sneedii is found in New Mexico and Texas, USA, and also in Chihuahua, Mexico. It has a conservation rating of Least Concern. It is a dwarf, densely clustering plant with cylindrical branches smothered in characteristic white spines. It is pincushion-like in appearance and is in some demand by collectors. Selected artificially propagated clones are available in international trade, as are seeds which dominate the CITES trade data. A number of *Escobaria* are treated as subspecies of *Escobaria sneedii* by some authors but only subsp. *sneedii* and subsp. *leei* are listed in Appendix I.

Trade data shown in this slide is the sum of both taxa, with approximately equal trade between the two species. No wild trade has been reported and trade is in live plants and seeds. The USA is the main exporter.

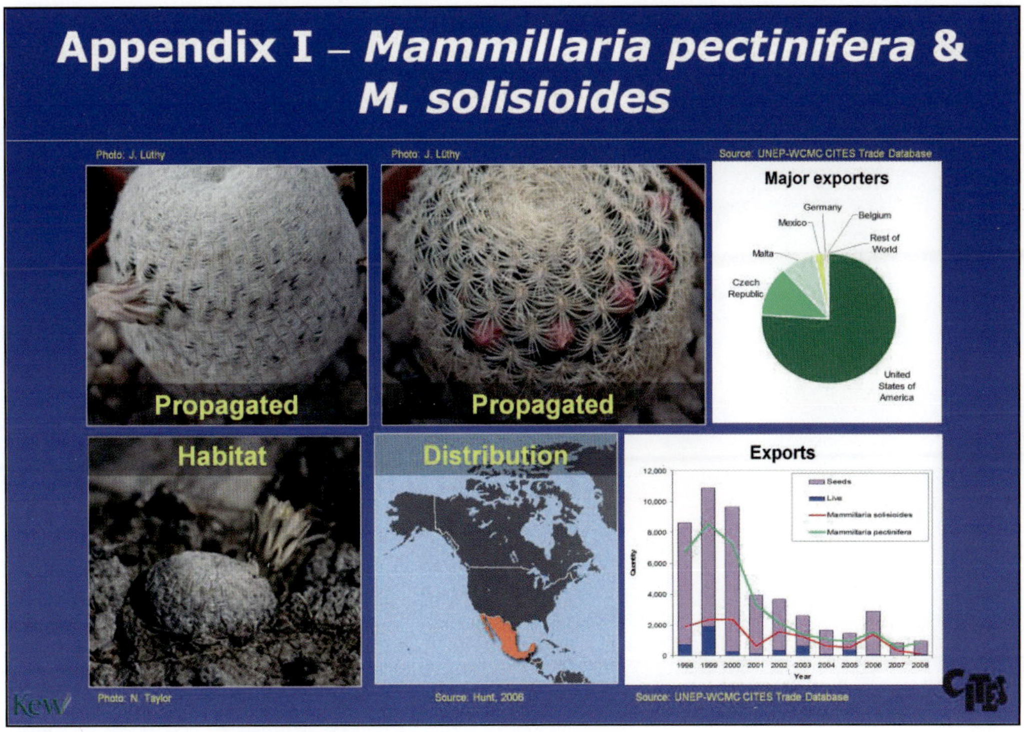

Slide 21: Appendix I – *Mammillaria pectinifera* & *M. solisioides*

The genus *Mammillaria* contains over 160 species concentrated in Mexico, but it is also found in south-western USA, the West Indies and South America. The great diversity in species and habit make this genus among the most traded cacti. Propagation by seed is easy for most species and there is a large number of artificially propagated plants in trade.

Two species are included in CITES Appendix I. *Mammillaria pectinifera* occurs in a small number of localities in the states of Puebla and Oaxaca, Mexico. It is a subspherical cactus with characteristic arrangement of white spines in a comb-like fashion over the stem. In nature the flat head of the stem barely protrudes from its gravel habitat and blends perfectly with its background. Despite this camouflage it has been collected to near extinction in some places for international trade.

Mammillaria solisioides is thought to be confined to some 6,000 plants in the states of Puebla and Oaxaca in Mexico. The plant has been in cultivation for some time and there is a good supply of mature propagated plants available.

Both of these species are slow-growing plants and historically there has been significant trade in wild plants. Seeds are now available and there is a good quantity of propagated plants in trade. There is still some wild collection and due to high commercial value there may be some illegal trade. The main threat is habitat loss through agriculture and expanding settlements.

Mammillaria pectinifera is the main species in trade, which is reported as artificially propagated and predominately in seeds with some live plants. The USA is the main exporter.

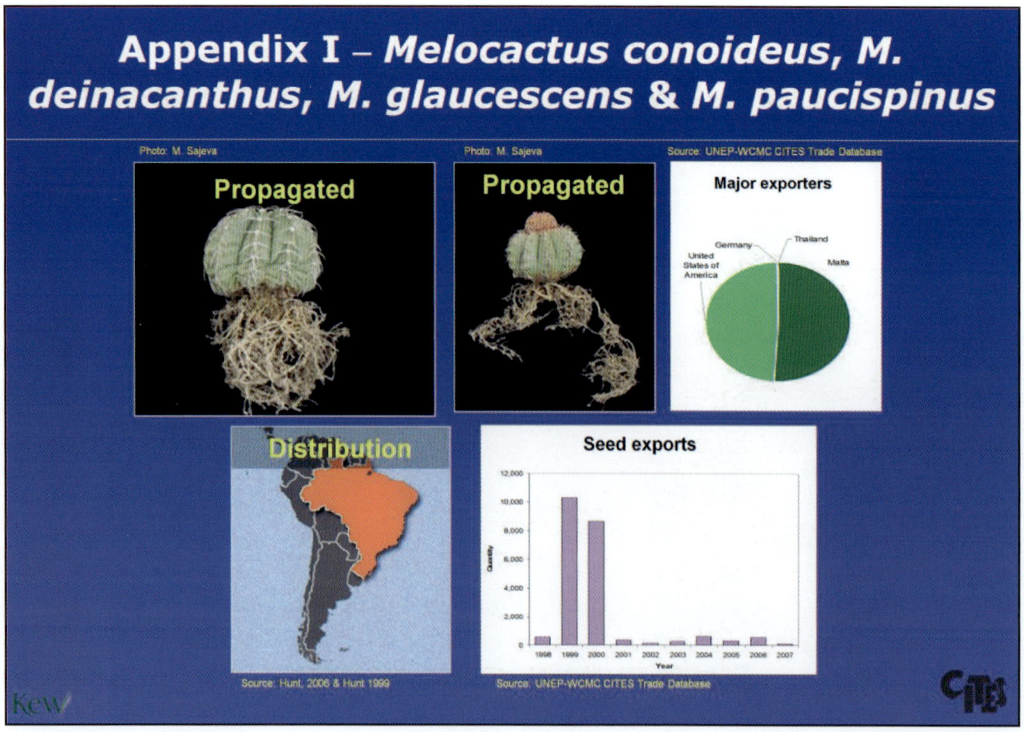

Slide 22: Appendix I – *Melocactus conoideus*, *M. deinacanthus*, *M. glaucescens* & *M. paucispinus*

The genus *Melocactus* includes 37 species distributed through the West Indies, southern Mexico and southwards into South America (especially Brazil). They are globular to short and cylindric plants and are solitary and occasionally clustering. The height is 0.15-1m and the diameter of the main body is 10-20cm. The body of the plant has 9 to 20 ribs usually fairly straight, with oval-shaped areoles up to 2.5cm, some of which with spines. The spines, usually stout, are often curved and pressed closely against the body and can be from 1.25-7.5cm long.

Mature plants have a distinctive hat-like bristly flowering zone (cephalium) at the top of the plant. This renders them of some interest to collectors although this is countered by the fact that they are more difficult to grow than more hardy cacti as they need sustained high temperatures in winter. Immature individuals may be particularly difficult to identify, as there is limited variation of stem and spine characters within the genus.

Four species of *Melocactus* are listed in CITES Appendix I. All of these species are endemic to Brazil.
Melocactus conoideus is endemic to a single mountain range and is Critically Endangered. Trade in wild specimens was reported in the 1980s but there is now limited interest in this species and artificially propagated stock is available.
Melocactus deinacanthus is Critically Endangered due to its restricted distribution and occurrence near roadsides. There is little known trade or collector interest in this species.
Melocactus glaucescens has a restricted distribution near urban areas and is Critically Endangered. Again trade in wild specimens was reported in the 1980s but there is now limited interest in this species and artificially propagated stock is available.
Melocactus paucispinus is known from some five widely scattered localities and is Endangered. However, there is no known recent trade in this species and there is little known demand. All of these taxa are now unlikely to be found in illegal trade.

All trade is reported as artificially propagated. *Melocactus conoideus* dominates the trade (97%) with the majority of trade as seeds.

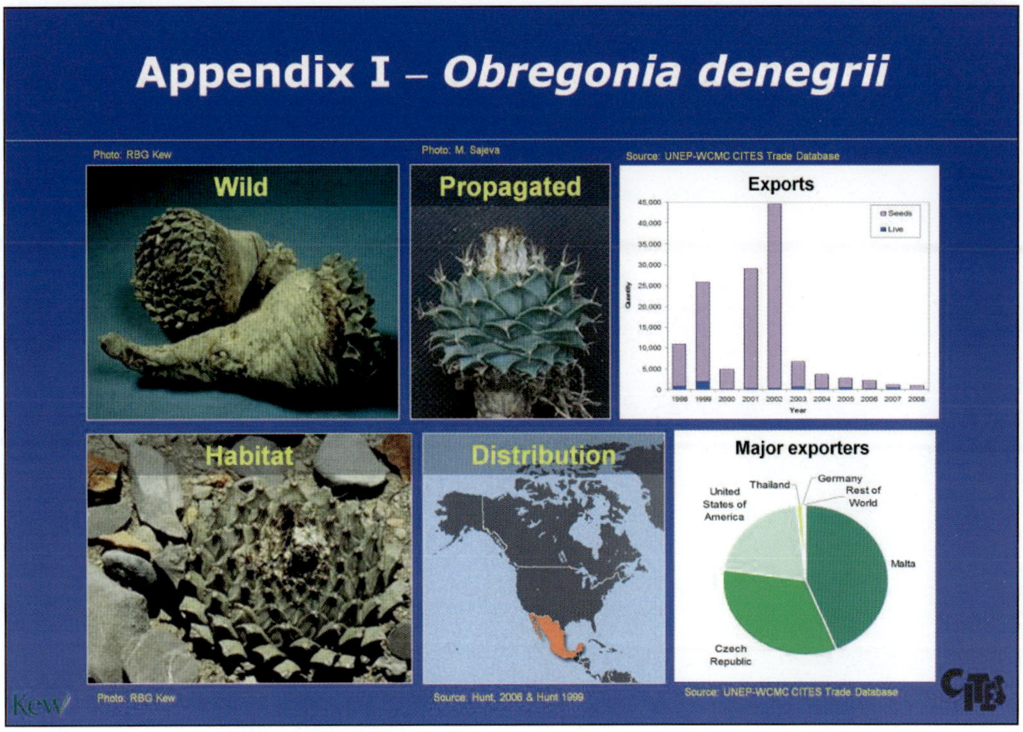

Slide 23: Appendix I – *Obregonia denegrii*

The genus *Obregonia* contains only one species and that is confined to a single valley in Mexico where it is now considered safe from extinction. It is a solitary, globular plant which can grow up to 15cm in diameter. This cactus is shaped like an artichoke head which gives rise to one of its common names – the artichoke cactus.

Obregonia denegrii has a conservation assessment of Vulnerable and has been heavily collected in its habitat, with extensive trade in wild-collected plants before it was listed on CITES Appendix I in 1981. Artificially propagated mature plants are now available and propagation from seed is quite easy. However, wild-collected plants are still attractive to some collectors and there may still be some in illegal trade.

Obregonia denegrii has been traded as live plants and seeds. Nearly all trade has been reported as artificially propagated, although some wild trade of live plants was recorded in 2000.

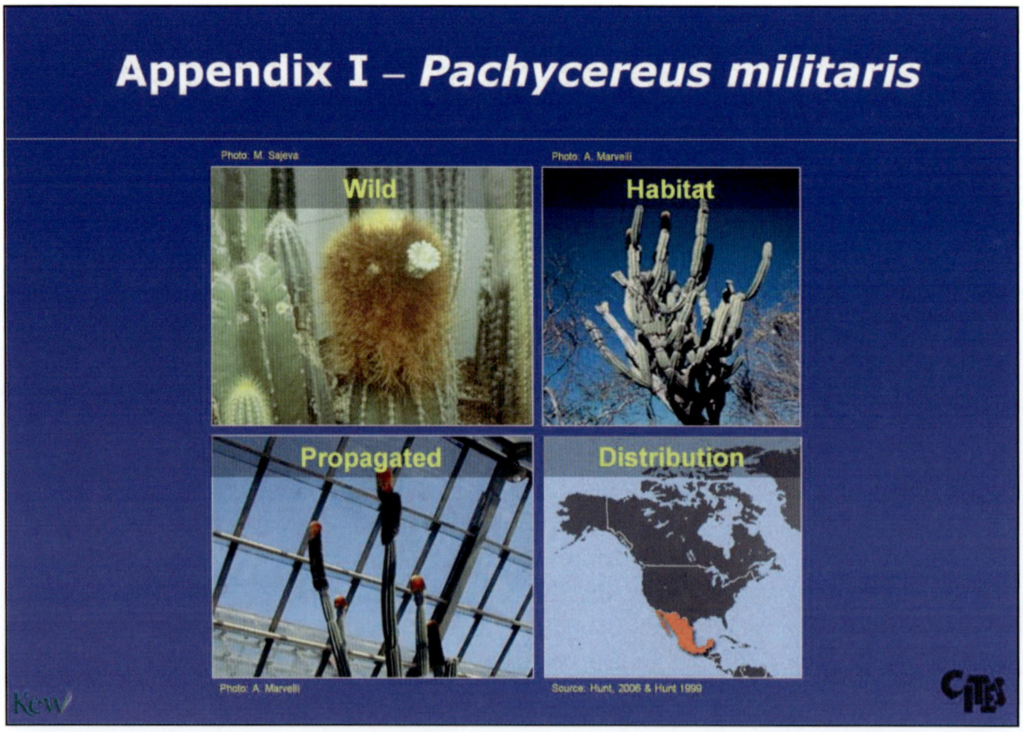

Slide 24: Appendix I – ***Pachycereus militaris* (*Backebergia militaris*)**

This is one of the 13 species of the genus *Pachycereus* and it is confined to Mexico (states of Colima, Guerrero, Jalisco and Michoacán). *Pachycereus militaris* grows as a tree which can reach 18m in height with numerous erect branches, dark grey-green in colour. The peculiarity of this species is to produce a distinctive hat-like red bristly flowering zone (cephalium) on mature branches.

The plant was subject to high levels of harvest in the 1970s and 1980s, with trees felled or branches hacked off to obtain the attractive cephalium. As a result, the species was listed in CITES Appendix I in 1983. Cephalium-bearing plants are not available from propagated material, so some wild stock may still be in trade. However, the peak demand for this 'novelty' plant has now passed and the species has a conservation status of Least Concern. Experience in growing the plant has now shown that such cephalia are of limited duration and thus of minimal horticultural value. They cease to grow as cuttings start to branch with time below the cephalium. The plant then directs all resources to the new shoots and the cephalium withers.

Trade is recorded as a small amount of live artificially propagated plants, and the Netherlands is the main exporter.

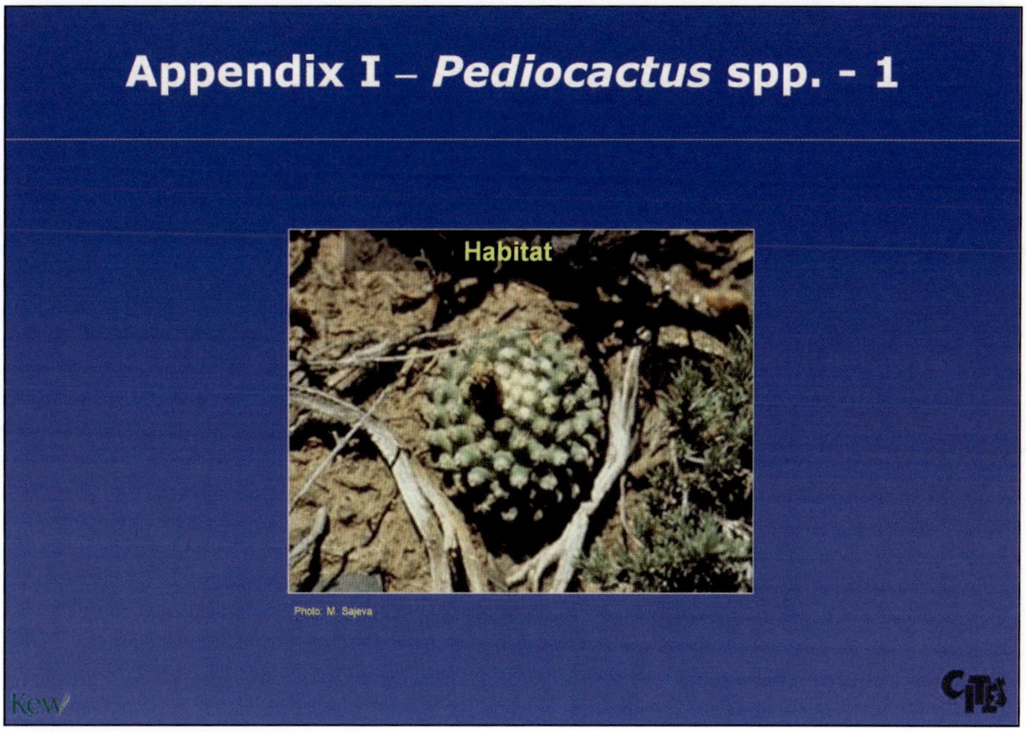

Slide 25: Appendix I – *Pediocactus* spp. - 1

The genus *Pediocactus* includes seven species, all of which are restricted in distribution to south-western USA. They are solitary or clustering, globular plants, from pale to dark green with a diameter of 2.5-15cm. Five species are listed in CITES Appendix I since 1983. Some of the Appendix I species are very distinctive and easily identifiable but others require some expertise to identify. Seeds are available and mature, artificially propagated plants are now available to the market. These are often grafted on frost tolerant stocks (*Opuntia* spp., *Echinocereus* spp.). Cultivation is not easy and most plants in trade are grafted, which is difficult to do with wild-collected plants and would not be profitable. The likelihood of illegal wild-collected plants entering the market is therefore not high.

Pediocactus bradyi and its three subspecies are dwarf cacti which blend into the gravel plains of Arizona and Utah and retreat below ground during summer, when dormant, making them difficult to find. The three subspecies each have a conservation rating of Vulnerable. There has been some interest by collectors in this plant as a curiosity or novelty item. It is difficult to propagate, and this is usually done by grafting. It is unlikely to be in illegal trade.

Pediocactus knowltonii is a dwarf cactus with a golf ball-like head. In habitat, it barely breaks the surface of its gravel substrate. It occurs in open woodland habitat on the borders of Colorado and New Mexico. It is one of the rarest cacti, currently rated as Vulnerable, and has been heavily collected since first described. However, it is now relatively common in cultivation and there is little or no demand for wild plants.

Pediocarpus paradinei is restricted to the watershed of the Colorado River in Northern Arizona and has a conservation rating of Vulnerable. Again, it is a species of limited interest to collectors. It is available to the market as propagated plants.

[Note to speaker: this slide shows Pediocactus knowltonii]

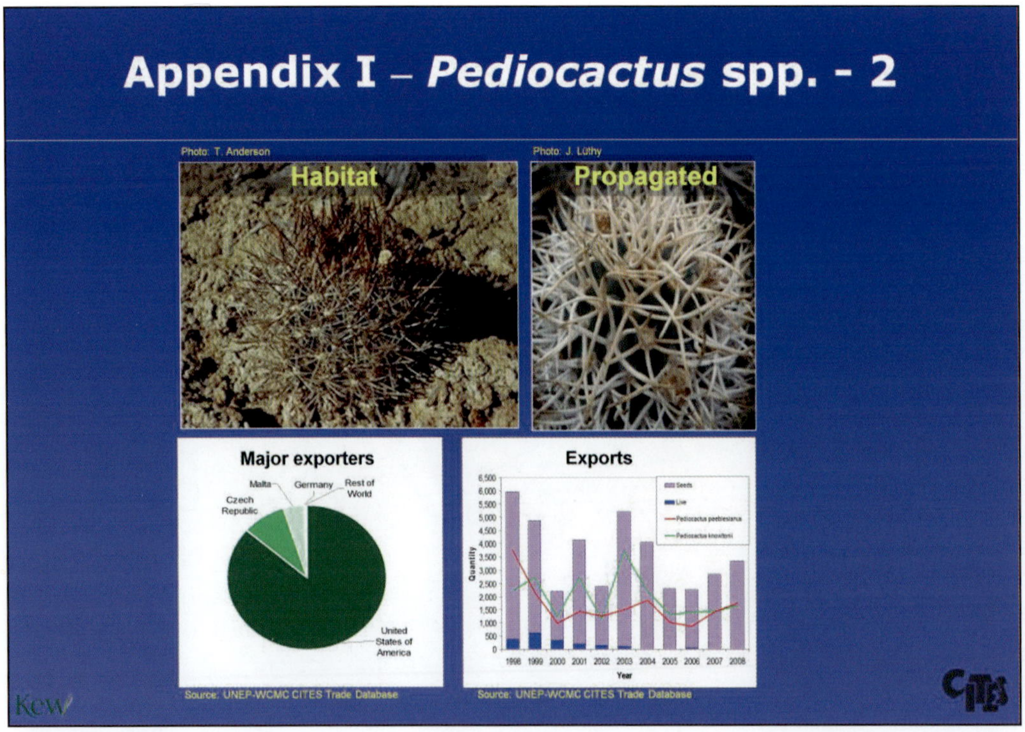

Slide 26: Appendix I – *Pediocactus* spp. - 2

Pediocactus peeblesianus, again Vulnerable, is a dwarf and difficult to find species of the Navajo Desert. It is of limited interest to collectors and is now available as grafted stock, which ensures better survival in cultivation.

Pediocactus sileri, rated Vulnerable, is known as the gypsum cactus due to its preferred substrate. It occurs in the borderlands of Arizona and Utah. It is very rare and difficult in cultivation and therefore is in international trade as seed, which may be wild and hence illegal in origin.

Trade is predominately in *Pediocactus knowltonii* (32%) and *P. peeblesianus* (26%). Trade data in this slide illustrates only *P. knowltonii* and *P. peeblesianus*, which are all recorded as artificially propagated seeds and live plants. Trade has decreased between 2008-2010.

[Note to speaker: this slide shows, top left: Pediocactus sileri, top right: P. peeblesianus]

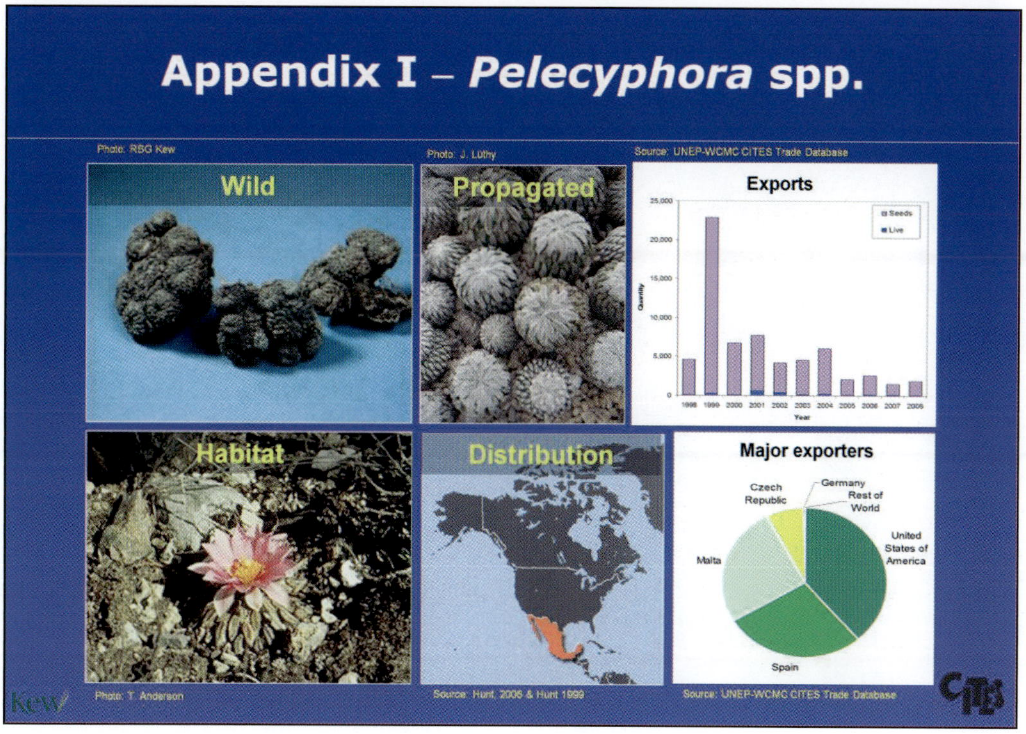

Slide 27: Appendix I – *Pelecyphora* spp.

This genus includes two species *Pelecyphora aselliformis* and *P. strobiliformis*. They are endemic to central and north Mexico. Plants are solitary or clustering. They are dwarf and characteristically very hard-textured, with elongated heads, 5-10cm in height and 2-6cm in diameter. In habitat the flattened head hardly emerges from the ground. The cactus heads are covered with hatchet-shaped spiny outgrowths (tubercles) which are spirally arranged. Species are slow-growing and therefore prone to illegal trade as mature wild plants.

Pelecyphora aselliformis is characterised by its spine cushions and arrangement of spines resembling woodlice. The head of the cactus looks as though it is swarming with these small creatures. Despite heavy collection in the past it is reported to be still frequent in habitat and has a conservation rating of Least Concern. It is easy to propagate, but is slow-growing.

Pelecyphora strobiliformis is often traded under its old botanical name *Encephalocarpus strobiliformis*. It has triangular tubercles and resembles a pine cone, hence its common name of the pine cone cactus. Highly sought after with a conservation rating of Least Concern, it is thought to be near extinct in its original site but is now known to occur in at least two other locations. It is available as artificially propagated plants from seed, but it is very slow-growing which has prompted illegal trade in mature wild plants.

Only two *Pelecyphora* species are recorded in CITES trade: *P. strobiliformis* (approximately 60%) and *P. aselliformis* (approximately 40%). Trade is recorded as artificially propagated and predominately in seeds with some live plants. Some wild trade was reported in 2000 and 2007.

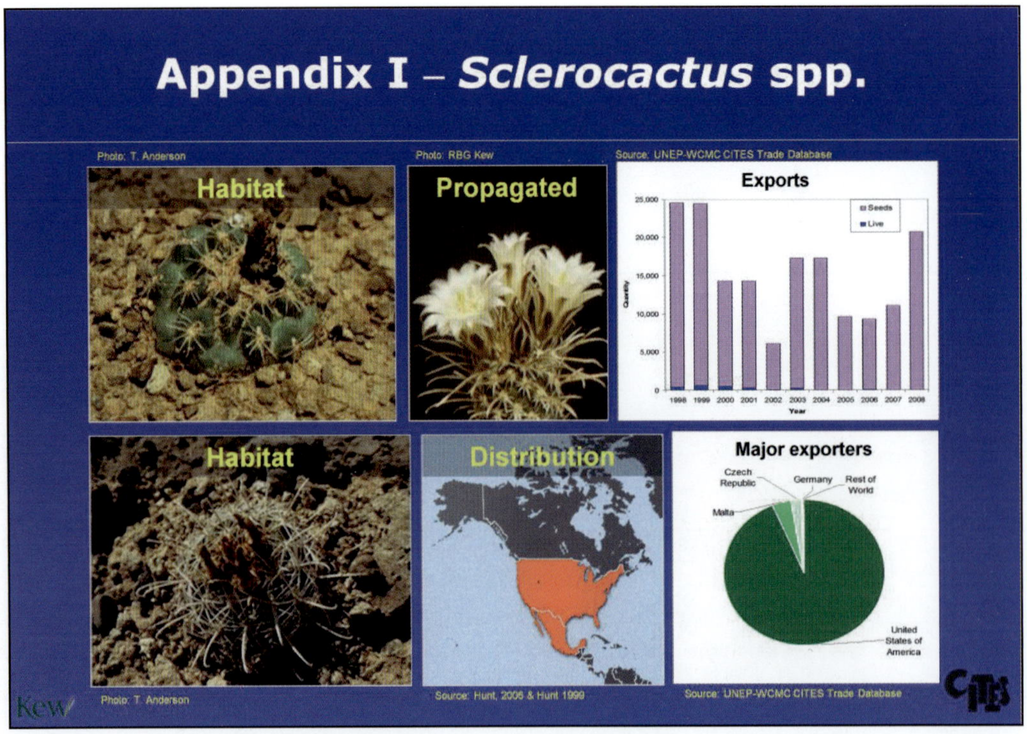

Slide 28: Appendix I – *Sclerocactus* spp.

The 20 species of the genus *Sclerocactus* originate from south-western USA and northern Mexico. The cylindrical plants are globular to short, up to 45cm tall and 10cm wide. Some species are dwarf and concealed in habitat, barely emerging from the ground and difficult to spot. The plant body ranges in colour from green to bluish-green. Most species have ferocious spines.

Eight taxa are included in CITES Appendix I since 1983 and one was added in 2003*. The others are listed on Appendix II. Trade in seeds is important in this group of Appendix I taxa. In general, the plants are difficult to cultivate and wild plants do not survive long in cultivation. Artificially propagated plants are usually grafted at seedling stage to produce more vigorous adult plants. Seed germination is also prone to be low. Therefore, there may be still some illegal wild collection of plants and seed for trade, especially in the case of *Sclerocactus mesae-verdae* which is rated Vulnerable.

All trade is recorded as artificially propagated and is mainly in seeds, with very few live plants. The CITES trade data is dominated by *S. papyracanthus.*

[**Note to speaker: these are Sclerocactus brevihamatus subsp. tobuschii, S. erectocentrus, S. glaucus, S. mariposensis, S. mesae-verdae, S. nyensis, S. papyracanthus, S. pubispinus* and *S. wrightiae.*]

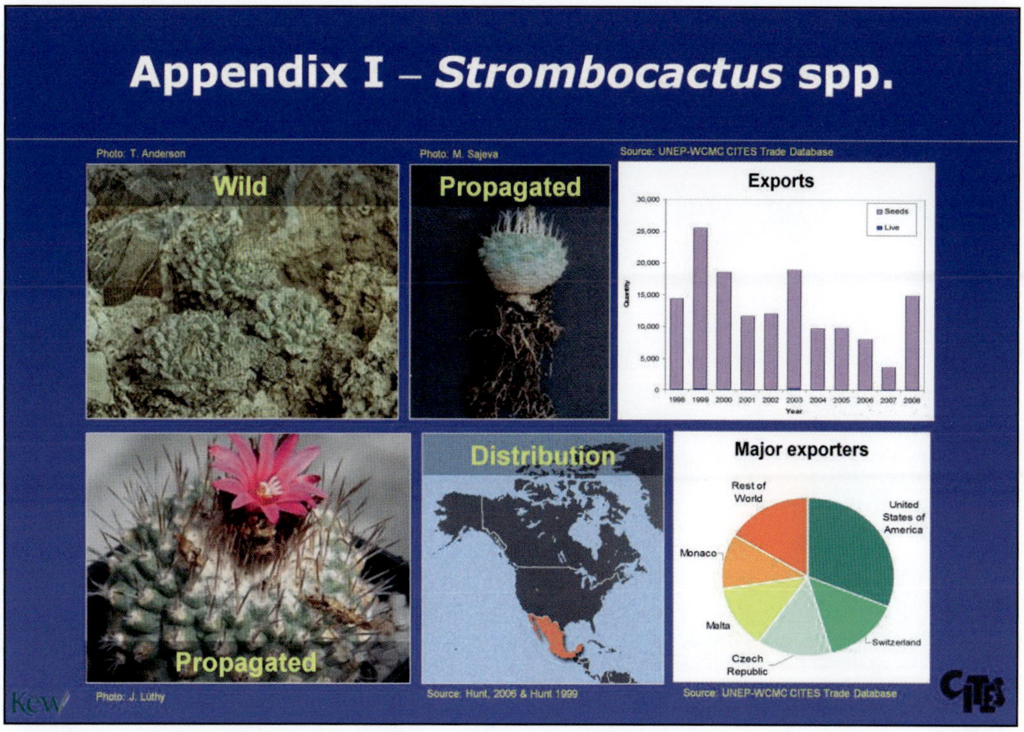

Slide 29: Appendix I – *Strombocactus* spp.

Strombocactus is a genus of one species (a 'monotypic' genus) which is confined to two states in Mexico, where it is now reported to be safe from extinction. The cactus has a stout taproot, is pine cone-like, very hard-textured, bluish-green to grey, up to 15cm in diameter with spirally arranged ribs with a characteristic fleck of wool at the centre of the plant.

Extremely slow-growing and very difficult to cultivate from seeds, it was in high volume trade from the wild in the 1970s and 1980s.

Strombocactus disciformis subsp. *esperanzae* is restricted in distribution and has a conservation rating of Vulnerable. First described in 1996, it was virtually immediately found in illegal trade in eastern Europe. Seeds are available but it takes years to grow from seed to mature, commercial-sized specimens.

There is high quality propagated material in trade but, due to the high value of wild specimens, illegal trade is very likely, especially in *Strombocactus disciformis* subsp. *esperanzae*. In habitat, spines are deciduous and wild-collected plants usually only have spines in the centre of the disc-shaped stem, whereas in cultivated plants, spines are much more permanent and usually cover the whole stem.

Trade data shown on this slide is only for *Strombocactus disciformis*. Trade is recorded as artificially propagated and is mainly in seeds, with some live plants. Some wild trade was recorded in 2000.

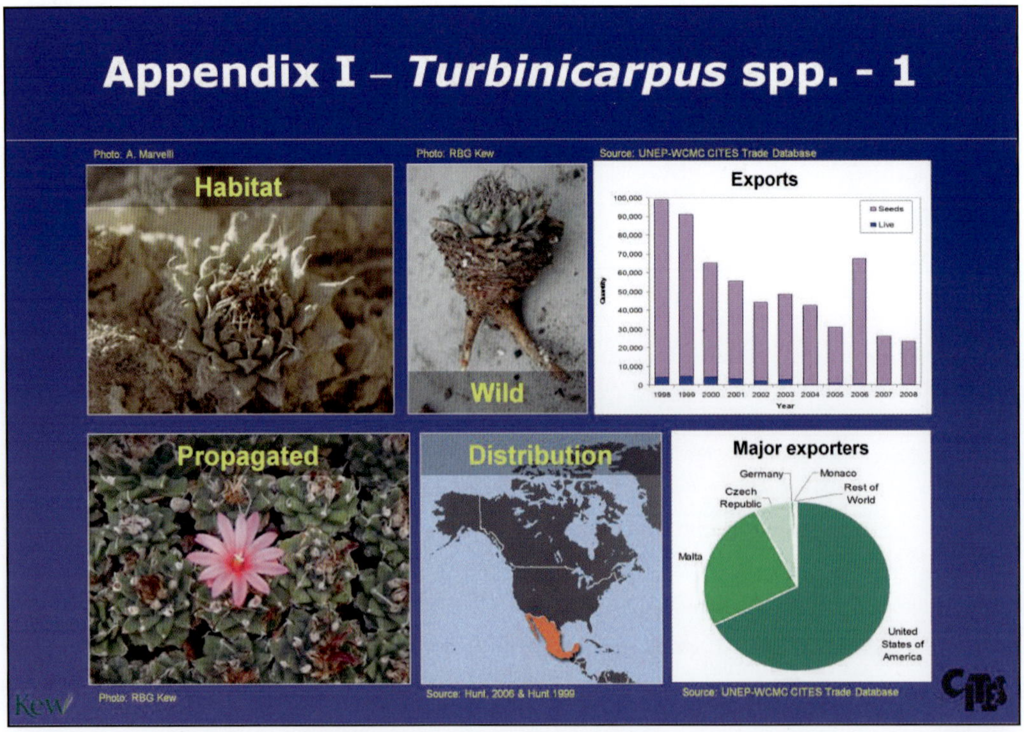

Slide 30: Appendix I – *Turbinicarpus* spp. - 1

The genus *Turbinicarpus* includes some 16 species and 20 subspecies distributed in Mexico. The full genus has been listed on CITES Appendix I since 1992. *Turbinicarpus* contains a significant number of local endemics which were very attractive to collectors when first discovered. There are 26 taxa threatened with extinction, 16 of which are Critically Endangered, making it perhaps the most threatened genus of cacti.

Large amounts of illegal collection occurred until very recently and this gave rise to significant illegal international trade. A few taxa still remain very difficult in cultivation such as *Turbinicarpus hoferi* and *Turbinicarpus mandragora* subsp. *mandragora*, both of which are Critically Endangered. However, in general, propagation is easy by seed.

CITES trade is dominated by *T. schmiedickeanus* subsp. *schwarzii*, *T. pseudomachrochele* and *T. valdezianus.* Trade is only recorded as artificially propagated and is mainly in seeds, with some live plants. Some wild exports were reported in 2000 and 2007.

Slide 31: Appendix I – *Turbinicarpus* spp. - 2

Turbinicarpus species are very diverse in habit and they are all in trade. *Turbinicarpus pseudopectinatus* and *T. valdezianus* (both Vulnerable) were heavily collected in the 1960s and 1970s.

In recent times, a number of new populations have been discovered and some plants were described as new subspecies or even species. This stimulated illegal trade in wild-collected plants, until demand could be satisfied with seedlings from artificial propagation. Plants were smuggled, most frequently, in personal luggage, sometimes in very large numbers. A number of populations were virtually wiped out by collectors, but later recovered when collection ceased.

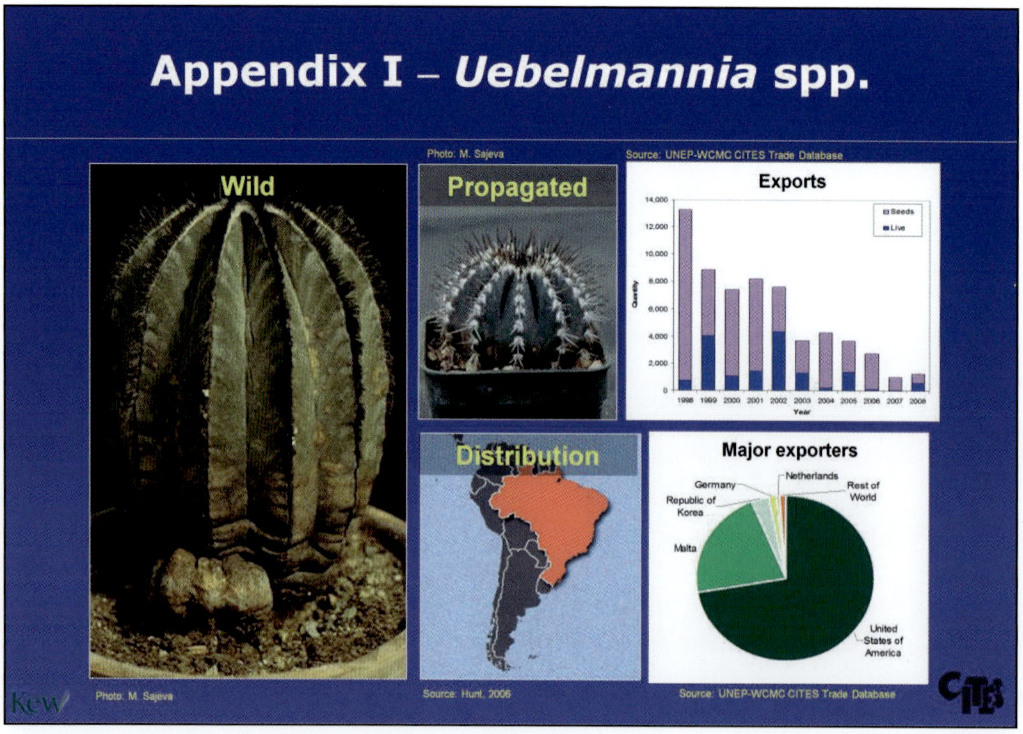

Slide 32: Appendix I – *Uebelmannia* spp.

The genus *Uebelmannia* contains three species and three subspecies, which are endemic (confined) to Brazil. Plants have mostly small, spherical to cylindric bodies. They are ribbed, spiny with a smooth or papillate (with blunt projections) epidermis that sometimes is covered in waxy scales. The stem has a shiny metallic aspect which is unique to the genus.

It is difficult to cultivate plants of the genus *Uebelmannia* and propagated plants are expensive. All species are usually grafted to encourage faster growth and better survival in cultivation. The peak of wild collection occurred in the 1970s and 1980s. However, there may still be wild collection and illegal trade, so inspection of imports should be carried out. Wild-collected plants have been found to carry lichens on the stem – a useful characteristic to check for, when inspecting plants.

Uebelmannia buiningii has been very heavily collected in the past both as plants and seeds. As a result it is considered to be one of the most threatened cacti of Brazil with a conservation rating of Critically Endangered.

Uebelmannia gummifera is a species of restricted distribution, and of limited availability as propagated plants and is rated Vulnerable.

Uebelmannia pectinifera (Least Concern) is a distinctive cacti, sharply and deeply ribbed in habit and is found on crystalline rocks on mountain slopes and plains where it remains relatively widely distributed. It is the most widely available *Uebelmannia* in cultivation.

Trade in *Uebelmannia* spp. is recorded as artificially propagated live plants and seeds, although export volumes have declined substantially since 1999. The most frequently recorded species in the CITES trade database are *Uebelmannia buiningii* and *U. pectinifera*.

Appendix II Cacti

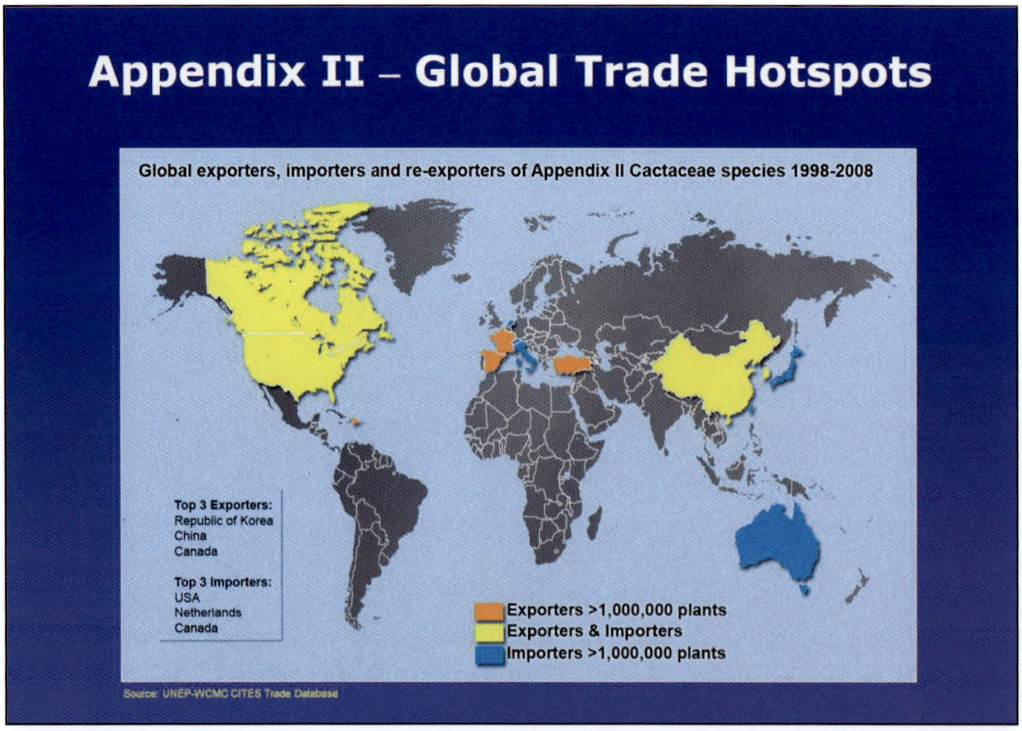

Slide 34: Appendix II – Global Trade Hotspots

This slide illustrates global exporters and importers of live, artificially propagated Appendix II cacti species between 1998-2008 (countries shown trade in more than 1,000,000 plants). Countries exporting cacti are illustrated in orange, those importing are illustrated in blue and countries that export and import cacti are illustrated in yellow. The top three exporters are the Republic of Korea, China and Canada. The top three importers are the United States, the Netherlands and Canada.

Trade quantity is measured by number of live plants, seeds and stems. This figure does not include data of trade in other parts and derivatives (e.g. carvings, fruit, extract, powder, roots etc) nor trade recorded in other units (e.g. kg, g, m³, m², shipments, sets, etc.).

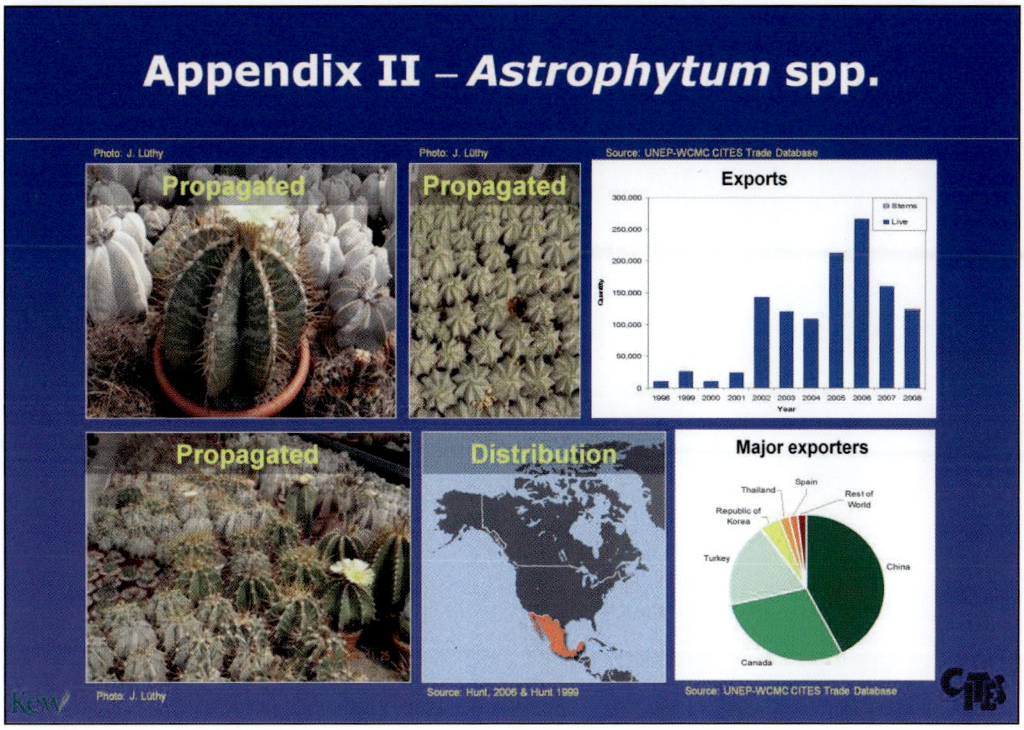

Slide 35: Appendix II – *Astrophytum* spp.

There are six species of *Astrophytum*, one of which, *Astrophytum asterias*, is listed on CITES Appendix I. All Appendix II species are found in Mexico. Four of the Appendix II taxa are not threatened and one, *Astrophytum caput-medusae*, is Endangered. Plants are usually solitary, rounded to columnar, and heavily ribbed giving in some taxa a 'bishops hat' appearance. There may be or may not be spines.

The CITES trade database records most trade in artificially propagated live plants with a small amount of wild trade. The majority of trade is in *A. myriostigma* and *A. ornatum*. Major exporters are China, Canada and Turkey.

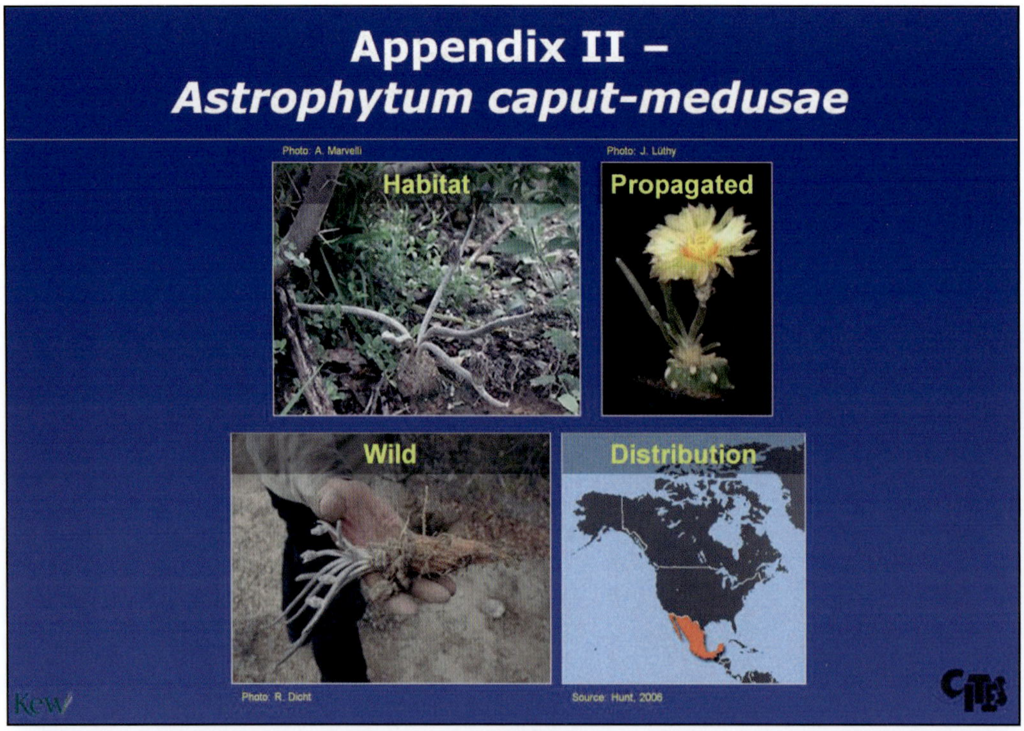

Slide 36: Appendix II – ***Astrophytum caput-medusae***

Astrophytum caput-medusae is one of the six species of the genus *Astrophytum*. It is found in Nuevo León, Mexico and has a conservation rating of Endangered. The main body of the cactus is buried in the ground with elongate short-lived wart-like projections or tubercles that arise from tuberous rootstock. Tubercles are cylindric and very slender, not bigger then 19cm x 2-5mm, of a grey-green colour. They are densely covered with minute white flecks. The vegetative tip of the tubercles has up to four rudimentary brownish spines some 2mm in length.

This species is, at the time of writing, the most recently discovered species from Mexico.

Seeds of Appendix II Mexican cacti exported from Mexico require CITES permits. To 2010 there is no recorded trade in the CITES trade database of this species. However, seeds and plants are now relatively widely available in the international market, apparently originating from illegally exported seeds. It is reported that the first seeds in the international market fetched prices of US$100 each.

The species has proved to be fast-growing and easy to cultivate and grafting of single tubercles has led to cultivated plants being available in the international market.

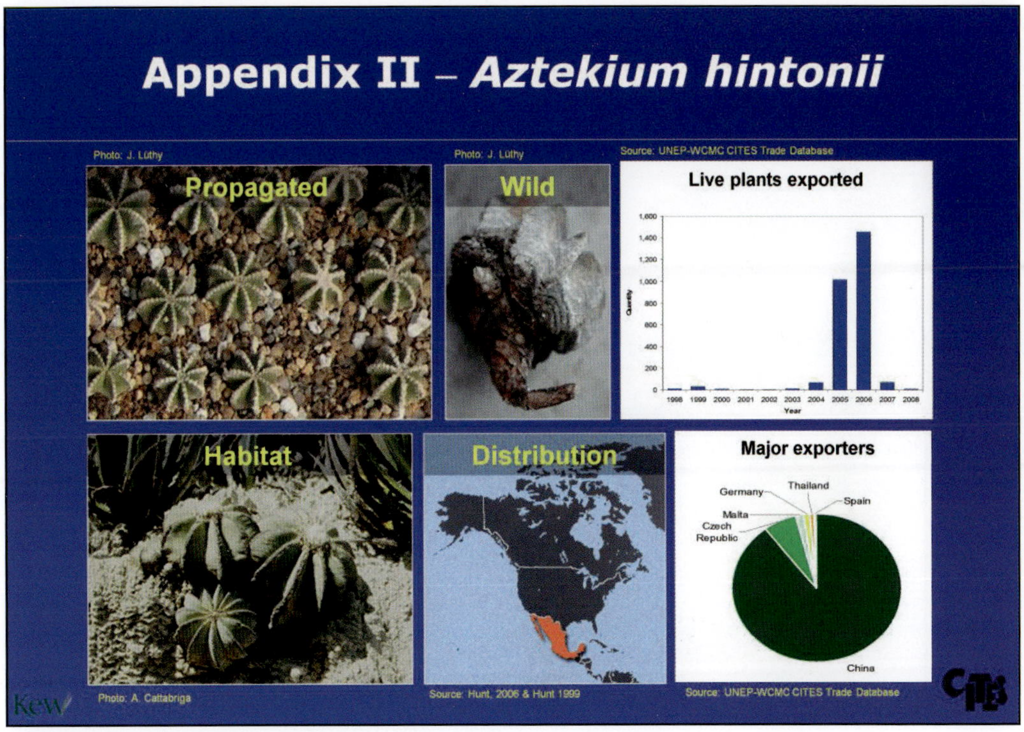

Slide 37: Appendix II – ***Aztekium hintonii***

This species is confined to Nuevo León, Mexico. It has a single, rounded stem that becomes slightly columnar with age. Some 9cm in height and 10cm wide, it is dull grey-green in colour with a sunken apex tipped with white wool. The surface is finely and continuously grooved by 10 or 15 ribs 0.6-1.2cm tall. The areoles are 0.12cm long and large and the epidermal cuticle is thick. Spines are from 0.3 to 1.3cm long, cream-white to dark-brown in colour, and are strongly curved and tortuous. It is described as a 'giant' when compared with the only other species in the genus.

Aztekium hintonii was discovered in 1991 on gypsum outcrops in Nuevo León and was immediately sought after by amateurs as a spectacular find. The species has a conservation rating of Least Concern and surveys have shown that populations reach into tens of millions. Seeds and cultivated specimens, mostly grafted, are available in the market. Growing it from seed has proved to be much easier than in the case of *Aztekium ritteri* (Appendix I), but growth is nearly as slow.

Aztekium hintonii has been illegally traded in large quantities and old, wild-collected specimens may be found today in some collections, offered on internet sites and in a few nurseries. Care should be taken when inspecting nurseries, shows and markets to check if such material is being offered for sale.

The CITES trade data records trade confined to artificially propagated specimens of live plants with the exception of 46 wild plants traded in 2000. The trade is totally dominated by exports from non-range States. China is the main exporter.

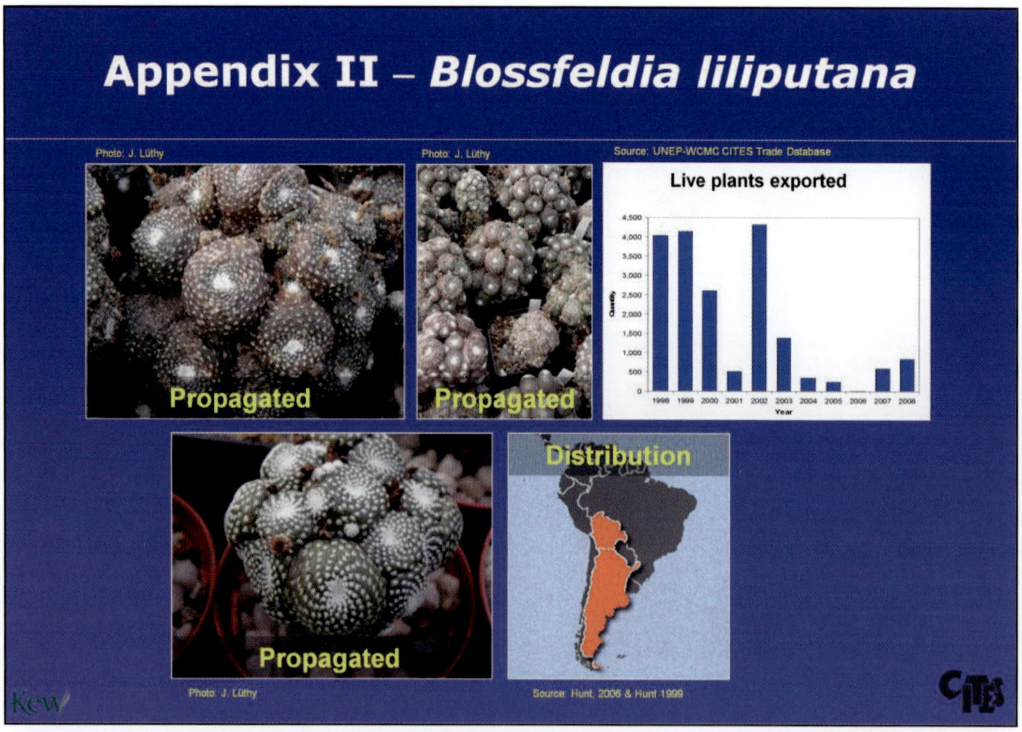

Slide 38: Appendix II – *Blossfeldia liliputana*

Blossfeldia liliputana is the only species included in this genus and it is found in Argentina and Bolivia where it is reported to be relatively widespread. It has a tiny button-like body with simple or densely tufted branches and is often found concealed in cracks in rock surface. It lacks ribs, tubercles (wart-like protuberances) and spines.

This is the smallest of all the cacti, only a few centimetres in girth when fully grown, hence its name. Seeds are available and propagation is not very easy because of the small size of seedlings. Large specimens (1-3cm in diameter!) should be checked to verify if of wild origin. It is of some interest to collectors due to its novelty.

The CITES trade data records trade only trade in artificially propagated plants with the exception of 40 wild plants exported from Peru in 2007. China is the main exporter and trade has increased between 2008 and 2011.

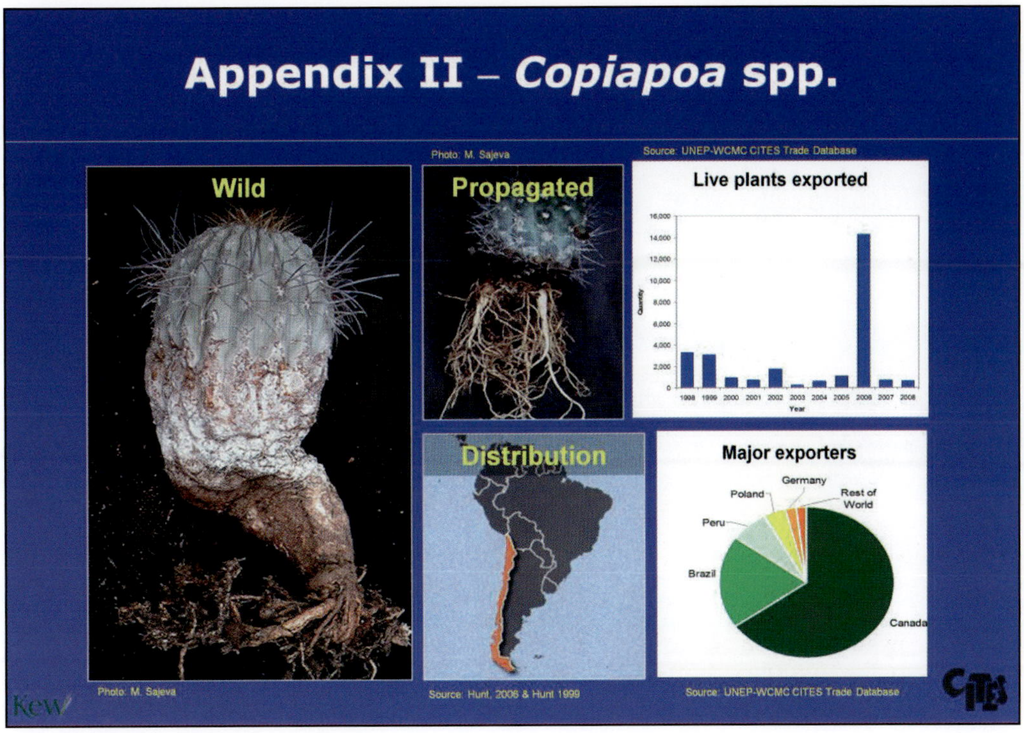

Slide 39: Appendix II – *Copiapoa* spp.

The genus *Copiapoa* includes 21 species and 9 subspecies which are confined to the coastal deserts of Chile, where species often survive on the seasonal fog, or 'camanchaca', which supplies the only moisture, while waiting for the once-a-decade rains. Eighteen taxa are threatened with extinction. Plants are globular to short columnar with age, often developing an enlarged taproot. They range from 8cm to 100cm in height and are grey to grey-green in colour. They are usually low-growing, mound-forming, spiny plants.

Wild plants are restricted from export for commercial purposes and the international trade is virtually confined to artificially propagated or claimed artificially propagated stock. Since CITES listing they are probably one of the most frequently smuggled genera of cacti from South America. The genus is attractive to collectors and has been found in illegal trade in some numbers. Illegal trade has routed through a number of countries bordering Chile.

Wild-collected plants may be held in cultivation for some time before shipping and then exported claimed as artificially propagated. Shipments of *Copiapoa* should be checked to verify if the plants are as indicated on export documents i.e. artificially propagated. Propagated plants are usually young, no more than 5 years old (up to 10cm in size depending on the species). Propagation is always by seed and the species are slow-growing. The milder conditions present in cultivation usually cause the plants to appear greener and less stout than those growing in the wild habitat.

The propagated plant in the slide is approximately 4 years old, and the wild plant could be over 25 years old. The wild plant in the slide was confiscated in Italy in 1984 in a shipment coming from Peru. Confiscations continue to take place in importing countries and imports from the region should be carefully checked.

The CITES trade data records trade which is virtually confined to artificially propagated live plants, with very low levels of wild trade reported. Major taxa in trade are *C. montana*, *C. cinerea* and *C. coquimbana*. Canada is the main exporter.

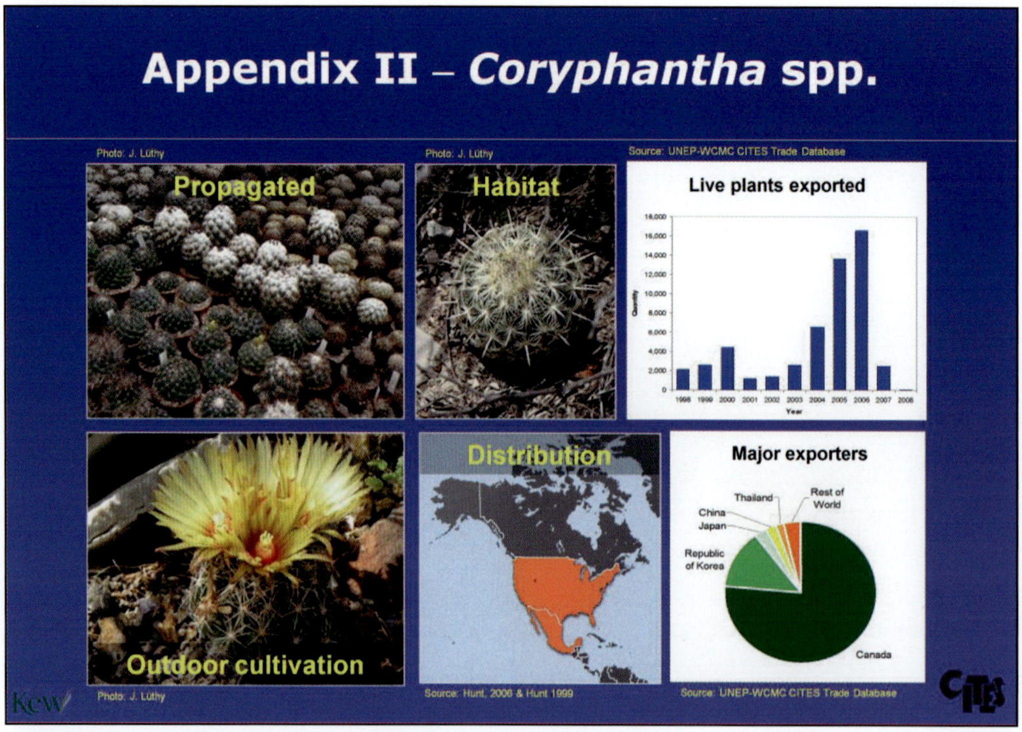

Slide 40: Appendix II – *Coryphantha* spp.

This genus contains 42 species and is one of the most widespread genera of North American cacti, occurring in the USA and Mexico. Eighteen taxa are considered to be threatened with extinction. *Coryphantha werdermannii* is included in CITES Appendix I. *Coryphantha* is a genus that consists of relatively small, spherical plants that have grooved tubercles instead of ribs. Cacti genera that are similar to *Coryphantha* include *Mammillaria* and *Escobaria*. Until recently there was some confusion over the boundaries between *Coryphantha* and related genera.

Coryphantha is relatively easy to propagate although they are slow-growing and some species may take ten years to flower. Despite this, they are frequent in cultivation and propagated plants are widely available. The genus is unlikely to be found in illegal international trade and there is limited reporting of confiscated plants in the last ten years.

The CITES trade data records trade as artificially propagated with only a very small quantity of wild-sourced trade recorded in 2002 and 2010. Trade data on this slide excludes data for Appendix I *Coryphantha werdermannii.*

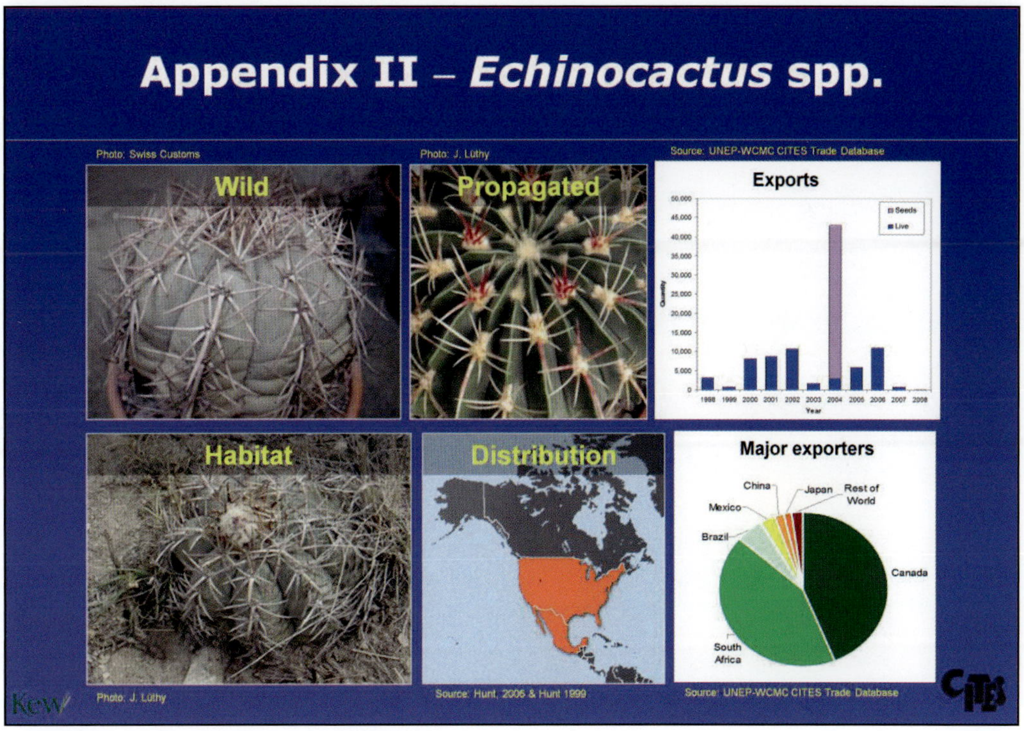

Slide 41: Appendix II – *Echinocactus* spp.

The genus *Echinocactus* comprises six species and one subspecies. It ranges from the south of Mexico to southern USA (Arizona, California, Nevada, New Mexico, Texas) occurring in desert or semi-desert habitat. Echinocacti or barrel cacti help shape the character of North American deserts. The genus is characteristically barrel-shaped with deep grooves and densely woolly crowns at the stem tips. One taxa is considered to be threatened with extinction.

With the exception of *E. grusonii* the genus is limited in cultivation and trade and with little likelihood of illegal trade. CITES trade data is dominated by artificially propagated plants. The slower growing species (*E. polycephalus*, *E. horizonthalonius* and *E. texensis*) are sometimes wild-collected.

CITES trade data does not include *Echinocactus grusonii*. This is illustrated on the following slide. Exports are predominantly in live plants, with a large quantity of seeds exported in 2004. Exports are mainly artificially propagated specimens, although some wild exports were recorded, most recently in 2010. Canada and South Africa are major exporters.

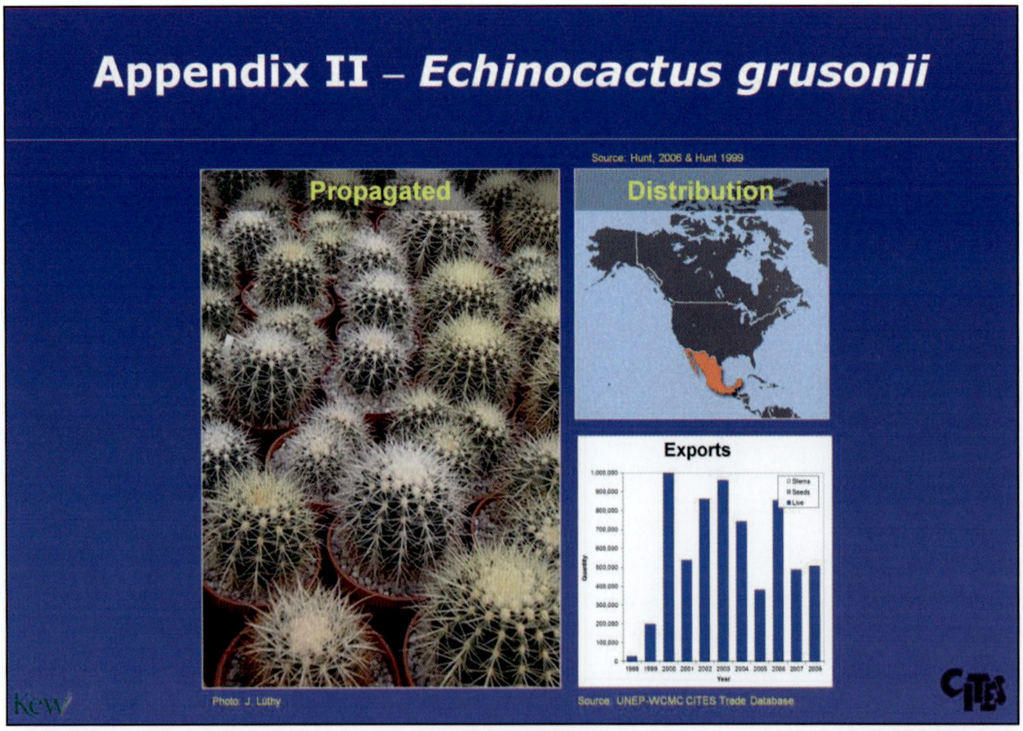

Slide 42: Appendix II – ***Echinocactus grusonii***

The golden barrel cactus is one of the most widely cultivated and traded cacti. Deeply grooved and covered in dense golden spines it can reach well over a metre in height and nearly a metre in diameter. Highly attractive, it is frequently used in landscaping and as a specimen pot plant.

Native to Mexico it is highly restricted in range and considered to be Critically Endangered. Originally thought to be confined to Querétaro and San Luis Potosí, where it occurred on steep hillsides and cliffs, most of its original habitat is now under the waters of the Zimapán dam. A striking plant, standing out it in its habitat, it caused great commotion in the cactus world when a new population was found in Zacatecas in 2005. The new site is some 500km from the other populations!

It is highly unlikely to be in illegal trade as a wild specimen. CITES trade data records exports that are nearly all in live artificially propagated plants, with small quantities of seeds and stems reported. Some wild-sourced exports were recorded in 2001 and 2010. China, Dominican Republic and Canada are the major exporters.

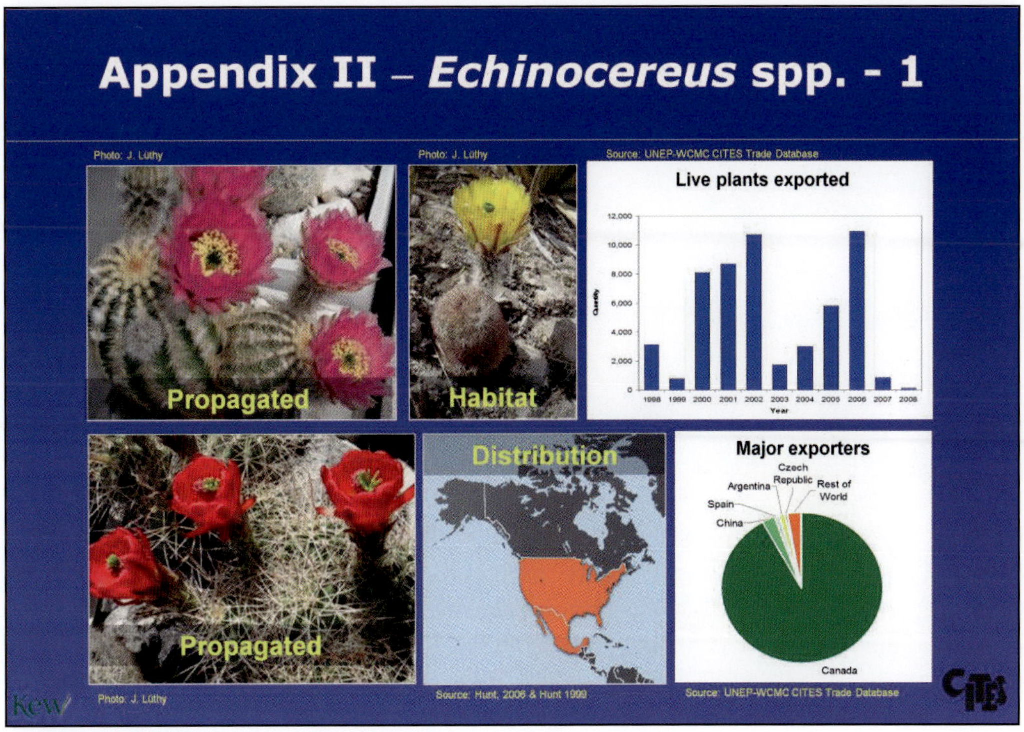

Slide 43: Appendix II – *Echinocereus* spp. - 1

The genus *Echinocereus* comprises 67 species and 39 subspecies and is one of the larger genera of North American cacti. It ranges from the south of Mexico to include the states of Dakota, California and Texas in the USA. It occurs in desert or semi-desert areas, up to 3000m above sea level. Two taxa are included in CITES Appendix I. Twenty five taxa are considered to be threatened with extinction.

Plants have a very variable appearance. They are frequently slow-growing plants which bear spectacular flowers, bursting through the skin surface. Usually they are low-growing, shrubby with fibrous or tuberous roots. They are tall, up to 60cm, and from 1-15cm wide. Branches are simple or clustering, spherical or cylindric, either self-supporting, trailing on the ground or rarely climbing like a vine. Lateral branches usually burst through the epidermis above an areole (felted disc).

[Note to Speaker: this slide shows, top left: Echinocereus reichenbachii (outdoor cultivation). Bottom left: E. mojavensis (outdoor cultivation). Top right: E. dasyacanthus]

Slide 44: Appendix II – *Echinocereus* spp. - 2

The genus contains some of the most popular North American cacti. The name has its origin from the Greek *echinos,* meaning hedgehog or sea urchin which is thought to refer to the spiny fruits, and *cereus* meaning candle.

Echinocereus is very easy to cultivate and a number of the northern taxa are cold hardy. Seeds are often sold as a 'winterhardy' mix with some capable of surviving northern European winters outside. Their small size, ease of cultivation and striking flowers make them one of the most popular cacti genera with both enthusiasts and amateur growers.

Widely available as cultivated material there is no little or no demand for wild species and illegal trade is therefore unlikely. CITES trade data is dominated by trade in artificially propagated plants with low levels of wild trade in 2001, 2002 and 2010. CITES recorded trade volumes are highest in *Echinocereus pectinatus*, *E. reichenbachii* and *E. engelmannii*. Canada is the main exporter.

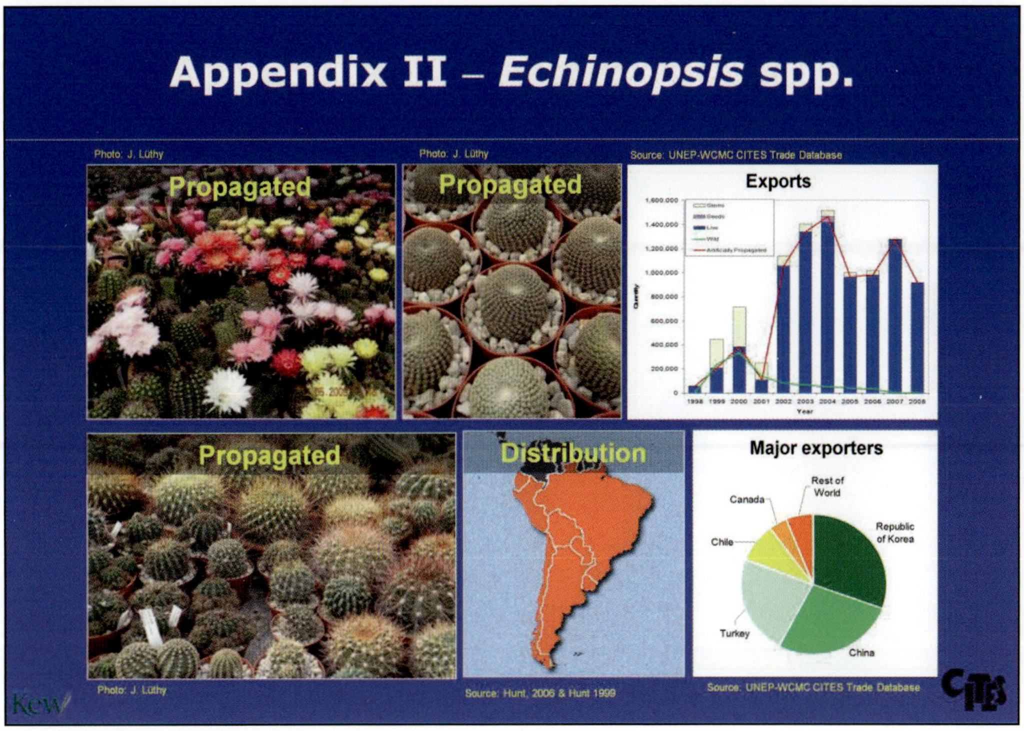

Slide 45: Appendix II – *Echinopsis* spp.

The genus *Echinopsis* includes 77 species and 24 subspecies distributed in Argentina, Bolivia, Brazil, Chile, Ecuador, Peru, Paraguay and Uruguay. They are shrubby, tree-like or columnar cacti ranging from 15-30cm to several metres in height and 40cm in diameter. The body of the cactus can have between 8 to 30 ribs which are usually straight and can be very prominent in larger species. The round white or grey to brown areoles (felted discs which bear spines) can be very close together. Each areole can bear 5-30 radial straight or curved spines. Spines are spreading and sometimes close pressed against the plant.

The species in trade are usually those which form the smaller, globular to short cylindric plants, whereas the species that form columnar, tall-growing plants are less common in trade. Propagation is easy for both these groups and plants are easily cultivated. Illegal trade is therefore not likely in this group.

Echinopsis chiloenensis is frequently found in trade as dried stem skeletons, cleaned and cut in sections. The spines are pushed in to form baffles, and they are filled with small pebbles or sand, creating a characteristic water flowing sound, to make an instrument called a 'rainstick'. Rainsticks are common in trade from Chile and Peru as novelty items. Frequently bought by tourists to Latin America, a special CITES exemption applies for limited personal imports.

The large volumes of wild trade is in stems for rainsticks. In addition to *Echinopsis chiloenenesis* – by far the dominant – *E. atacamensis, E. cuzcoensis, E. peruviana,* and *E. pasacana* have also been reported in trade as rainsticks. Trade is largely in live artificially propagated plants. Such exports are predominantly in *E. chamaecereus and E. silvestrii.* The Republic of Korea, China and Turkey are major exporters.

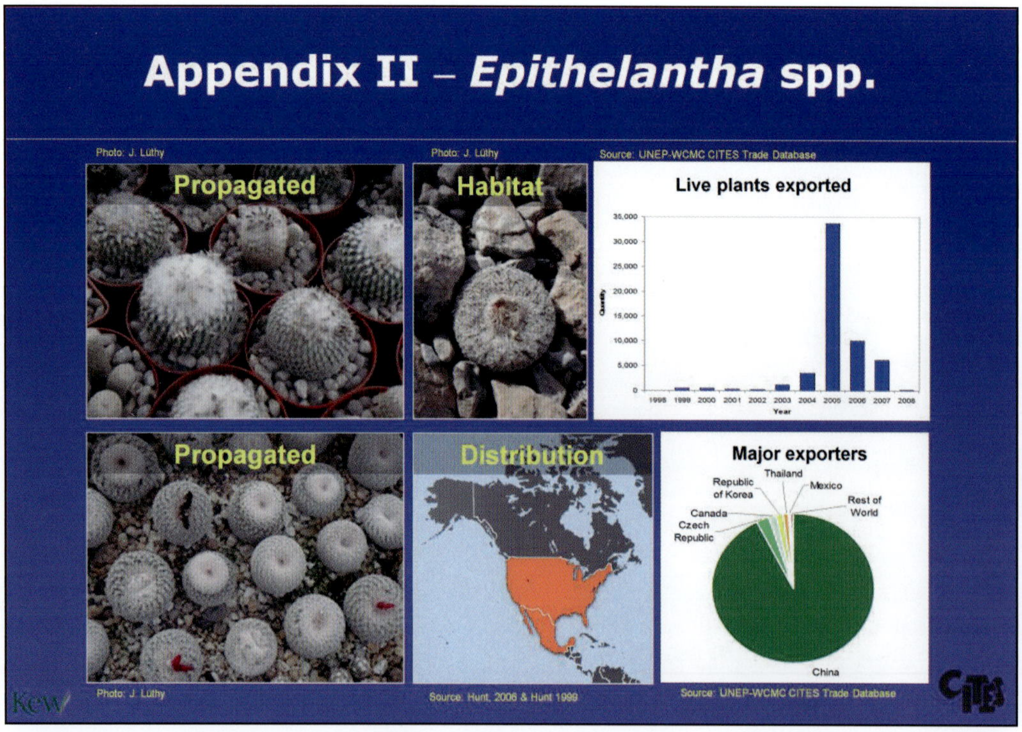

Slide 46: Appendix II – *Epithelantha* spp.

The genus *Epithelantha* (button cacti) contains two species from northern Mexico and south-western USA. Button cacti are solitary or clustering and some forms may have a taproot. Individual heads are 2.5-5cm in diameter and are covered with masses of minute tubercles and even smaller areoles. The areoles bear up to 40 very short, white or brownish comb-like spines.

In the past there has been wild collection of plants of this genus, but today you are unlikely to find wild-collected specimens in trade. Seeds are available and germinate and grow easily, although they are slow-growing.

Epithelantha bokei is restricted to the Big Bend region of west Texas and across the border in Mexico and is rated Vulnerable. *Epithelantha micromeris* is more widespread in nature and cultivation, in the wild occurring from eastern Arizona, New Mexico, west Texas and into Coahuila, San Luis Potosí and Nuevo León in Mexico. It has a conservation rating of Least Concern.

The CITES trade data records trade as artificially propagated, live plants, mainly exported from China, with the exception of six wild-sourced plants from Japan in 2010.

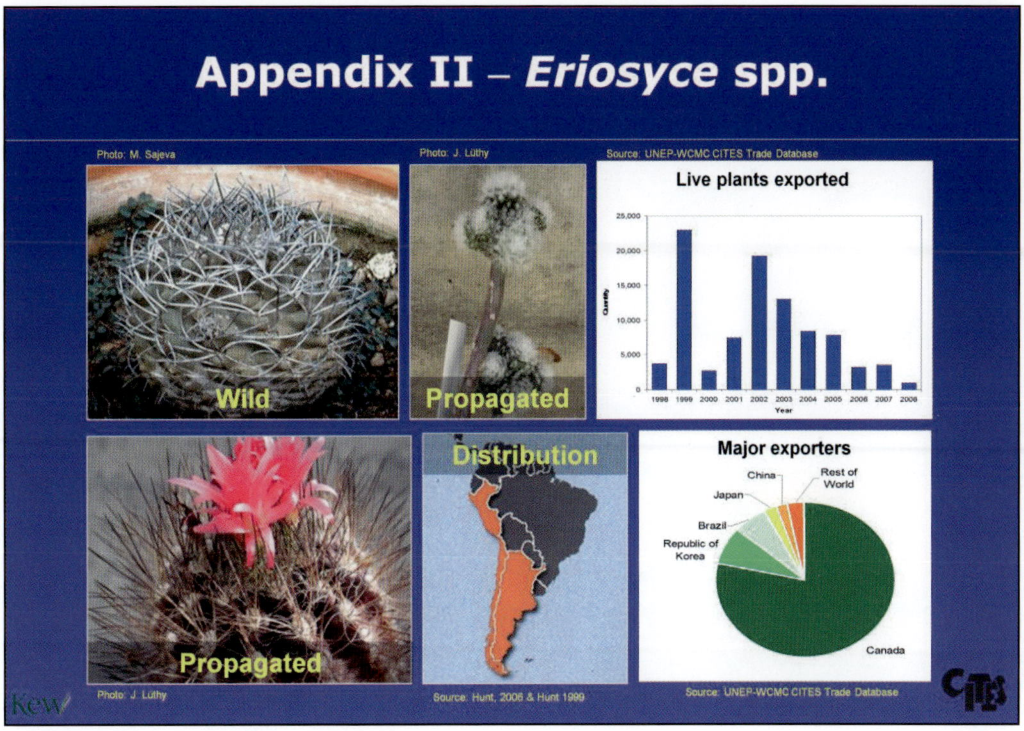

Slide 47: Appendix II – *Eriosyce* spp.

The genus *Eriosyce* contains 32 species and 19 subspecies found in Argentina, Chile and Peru. They occur from sea level to 300m on the eastern and western slopes of the Andes reaching into Argentina. Plants are usually simple, small and spherical or barrel-like. They have from 5 to 70 ribs and the spines are few to numerous, sometimes bristle-like. They are pollinated by insects and humming birds.

Plants of this genus can be propagated by seed, but currently there is not a high demand for this genus. In the past there has been some trade in illegal wild-collected specimens.

The CITES recorded trade data is dominated by trade in artificially propagated live plants with limited wild trade in 1998, 2002, 2004 and 2007. Canada is the main exporter.

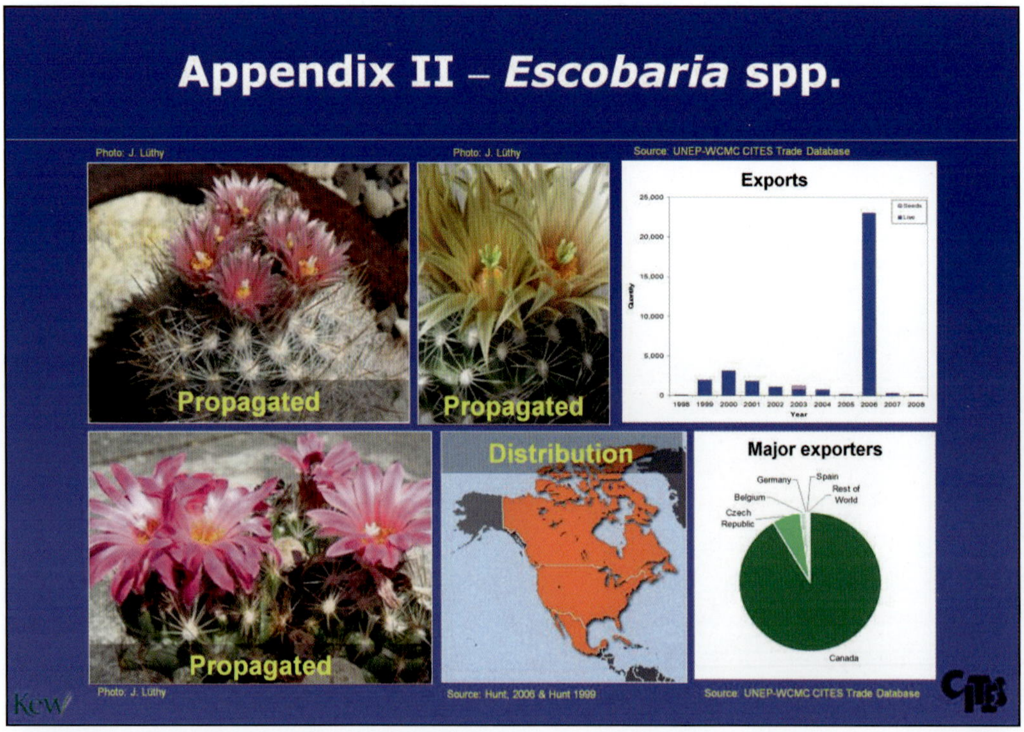

Slide 48: Appendix II – *Escobaria* spp.

The genus *Escobaria* contains 19 species and 4 subspecies. Two species are listed on CITES Appendix I. Closely related to *Mammillaria* the genus is distributed from southern Canada through western USA extending to Mexico and just reaching Cuba.

The genus *Escobaria* includes species of small, low-growing, spherical to cylindrical, solitary or clumping stems with tubercles and no ribs. Spines are usually short and fine, densely covering the plant. Flowers are typically pink. Many of the species were once included in *Coryphantha* and earlier in *Mammillaria*. *Escobaria* differs from *Mammillaria* in that it flowers from new growth at the top and lacks dimorphic areoles – that is the areoles do not have separate parts for producing spines and areoles.

A number of species are common in cultivation, and although cultivation may be difficult, the northern species are cold hardy and therefore attractive to gardeners. All CITES trade data records are in artificially propagated live plants and seeds. Wild specimens are not sought after and illegal trade is unlikely. Trade has decreased substantially between 2006-2011.

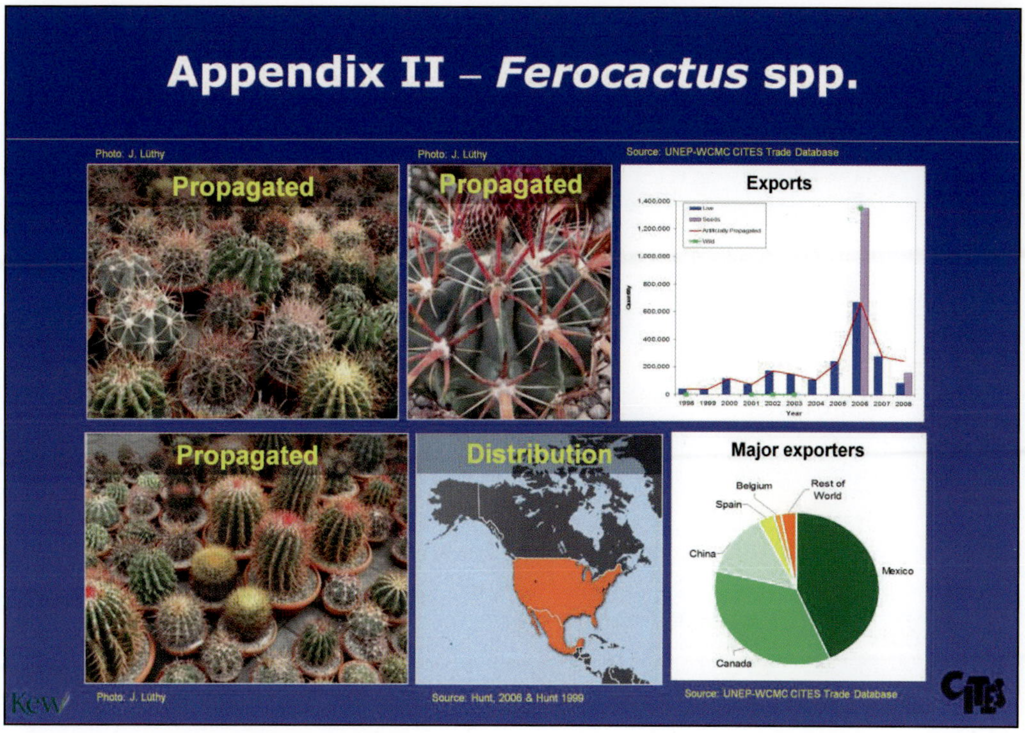

Slide 49: Appendix II – *Ferocactus* spp.

The genus *Ferocactus* includes 28 species and 14 subspecies distributed in the south-western USA, in a number of Mexican states and stretching around the Gulf of California. Species range from small plants that are globular or more or less cylindrical in habit to tall specimens that reach up to 3m in height and 1m in diameter. Branches, when present, are rounded or cylindrical. The stems have prominent ribs which in young plants appear as tubercles on the stems. Areolas (felt discs) are usually quite large with nectar-secreting glands and bear 10-20 radial spines that are often long and threatingly 'fierce', hence its Latin name based on the Latin *ferox* meaning fierce.

These barrel cacti are very attractive plants and the larger specimens are often used for landscaping in temperate climate areas or for decorating hotels and office blocks. Propagation is easy and wild-collected plants are seldom seen in the market places. Occasionally large specimens are removed from the wild during road construction and are sold with official permits and identification tags for landscaping. International illegal trade is unlikely.

CITES trade data is virtually confined to artificially propagated live plants, with a small amount of wild exports, notably a large export of wild seed from Mexico in 2006. *F. latispinus* is most frequently seen in trade. Mexico, Canada and China are the main exporters.

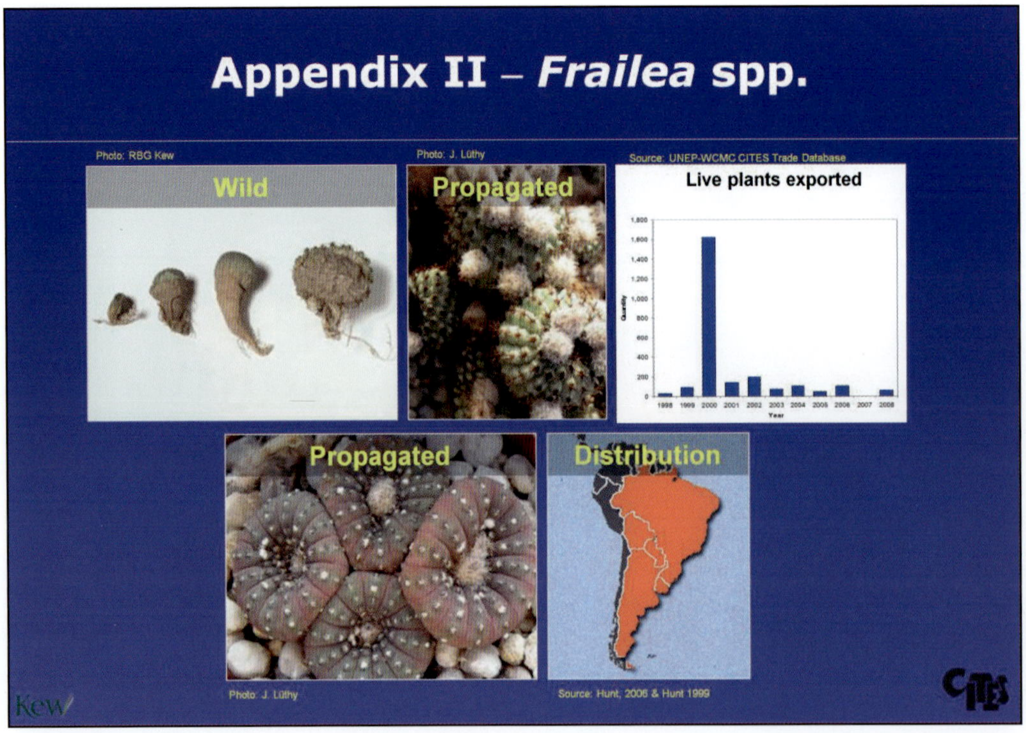

Slide 50: Appendix II – *Frailea* spp.

This genus includes 12 species and 6 subspecies spreading widely from Bolivia to Uruguay, Brazil, Argentina, Paraguay and possibly Colombia. The plants are small and solitary, globular to short cylindric. The branches are rounded to cylindric and are usually weakly ribbed or tuberculate. The average diameter of an individual head is 2.5-3.5cm. The colour of the body ranges from light-green to reddish-brown or even violet-brown. The small areoles bear 8 to 20 short, fine radial spines.

These small plants are very fashionable and specimens of wild origin were found on the international market. However propagation by seed is easy and nowadays artificially propagated specimens are widely available. The CITES trade data is confined to trade in artificially propagated live plants, with the exception of a small amount of wild-sourced plants from Japan in 2010. The main species in trade is *Frailea castanea*. Wild specimens were present in illegal international imports by mail intercepted in Poland and the Czech Republic. The small size of mature plants make it easy to hide them in cigarette or video cassette boxes. Canada is the main exporter.

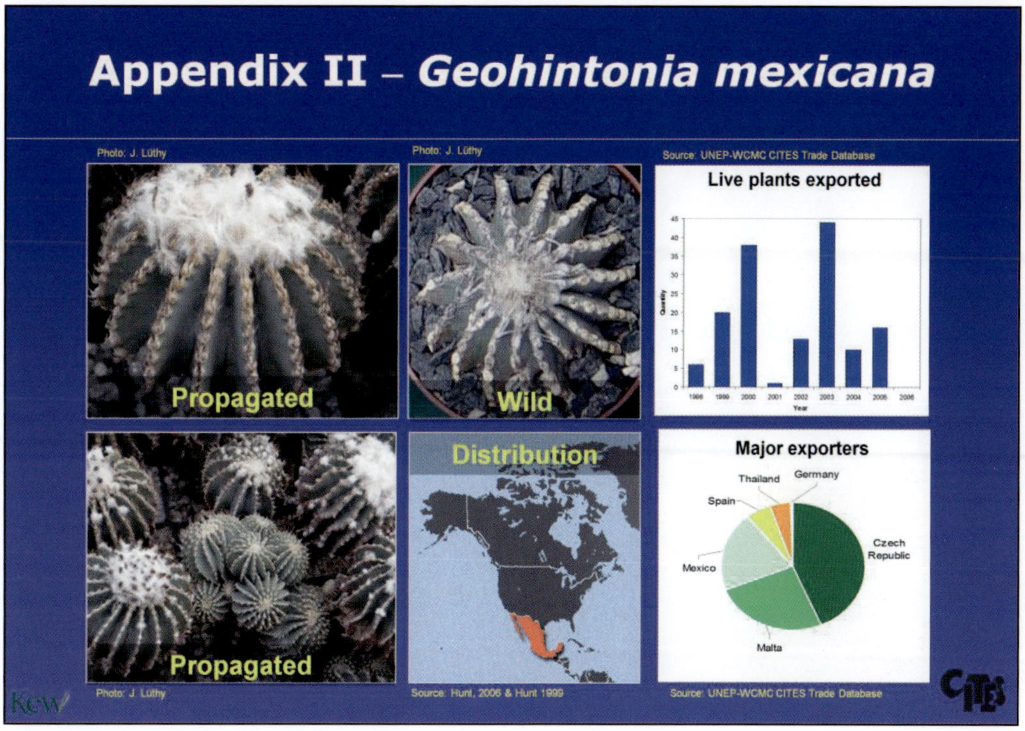

Slide 51: Appendix II – *Geohintonia mexicana*

Geohintonia is a monotypic genus confined to Mexico, where it is restricted to gypsum cliffs and hillsides in the Sierra Madre Oriental in Nuevo León. The plant's body, usually about 10-15cm by 10-14cm, is spherical, eventually short cylindrical and ribbed with woolly apex and bluish grey-green epidermis. The few spines are fragile and soon deciduous.

Discovered in 1992, it was immediately sought after by amateurs as a spectacular new taxa to their collections – a new genus of cactus. Seeds and cultivated specimens, mostly grafted, are available on the market. Care should be taken when inspecting nurseries as there has been significant illegal trade in wild specimens.

The CITES trade data shows only a small amount of trade in artificially propagated specimens with the exception of limited wild trade in 2000 and 2010. The Czech Republic, Malta and Mexico are the main exporters.

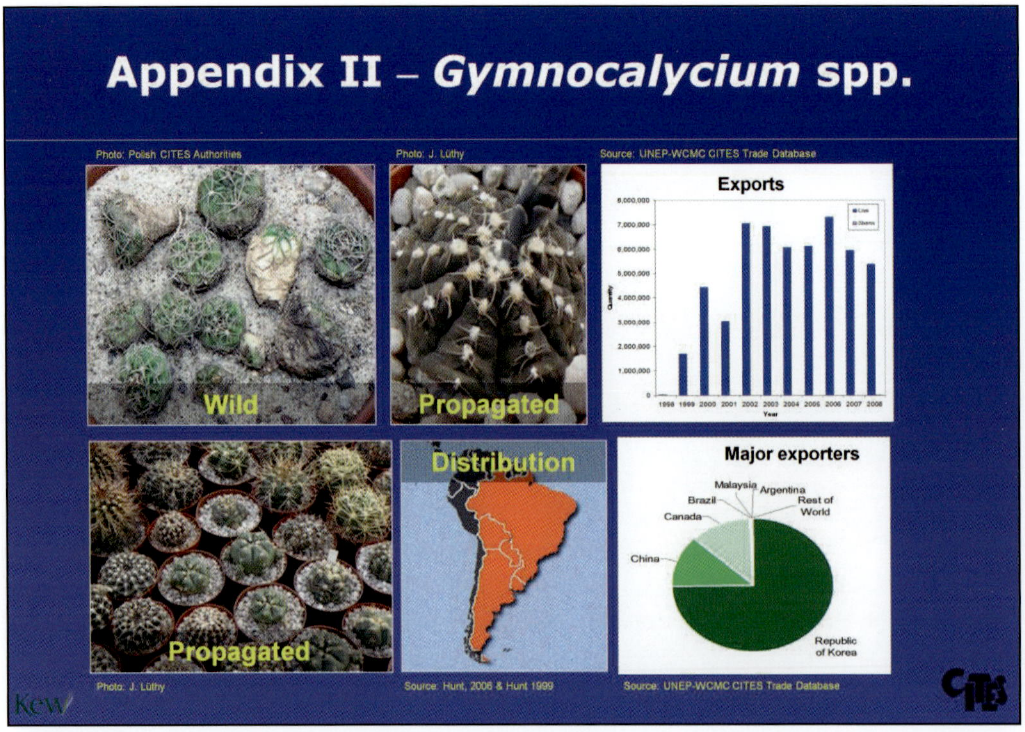

Slide 52: Appendix II – *Gymnocalycium* spp.

The genus *Gymnocalycium* includes 49 species and 14 subspecies that come from Argentina, Bolivia, Brazil, Paraguay and Uruguay, occurring east of the Andes. Plants are generally low-growing and are very variable in habit with bodies going from depressed to spherical or short-cylindric. Their colour can vary from green to reddish-brown and greyish-green. Most of the species are unbranched but some are clustering. The ribs are distinctly prominent and range in number from 4 to 15 or more, sometimes spiralling, variously tuberculate and spiny.

The genus is very common in cultivation and among the most popular for enthusiasts. Most specimens are propagated from seed. Some species, such as *Gymnocalycium spegazzinii* and *G. saglionis,* were wild-collected and entered international trade illegally. It is still possible that some illegal trade occurs but it is unlikely to be significant.

The CITES trade data shows that the trade is virtually confined to artificially propagated plants, with low levels of wild trade recorded. The majority of trade is in *G. mihanovichii.* The main exporters are China, Canada and the Republic of Korea.

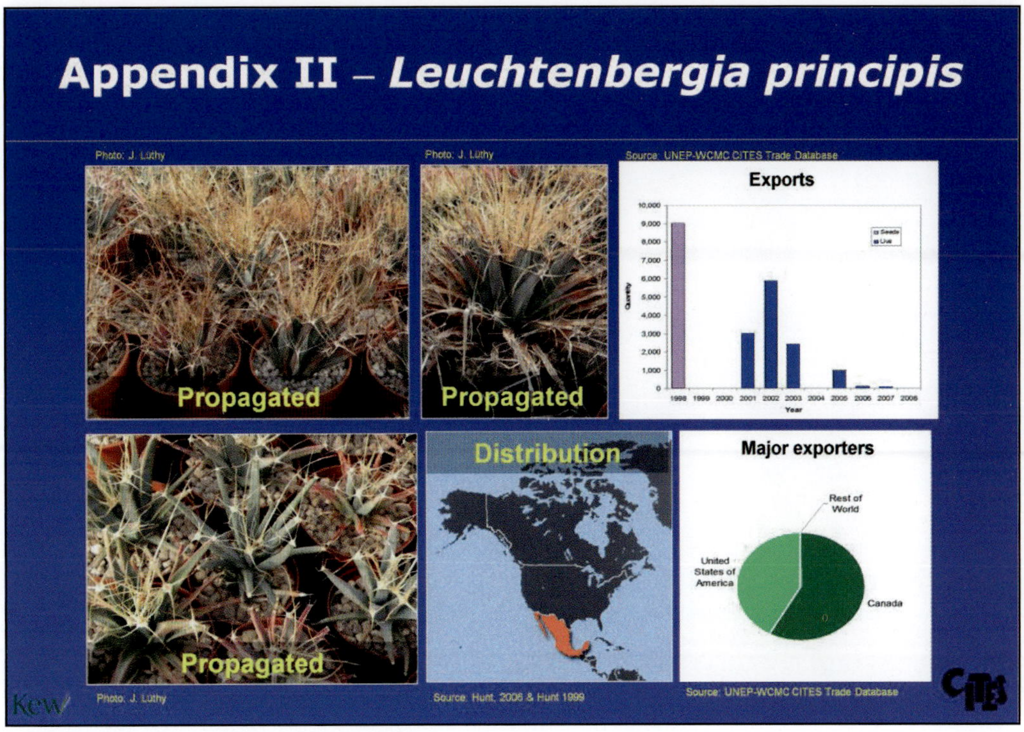

Slide 53: Appendix II – *Leuchtenbergia principis*

Leuchtenbergia, or agave cactus, is a monotypic genus confined to Mexico. It is reported to be widespread but rare from Cohauila to Hildalgo in the Chihuahuan Desert.

Leuchtenbergia is popular in cultivation for its long-lasting attractive yellow flowers and is hardy in cultivation. Solitary or slowly offsetting it is very slow growing but can eventually grow up to 70cm high, with a cylindrical stem which becomes bare and corky at the base with age. It has long, slender, bluish-green triangular tubercles 6-15cm long, with purplish-red blotches at their tips. The tubercles are topped with papery spines, making the plant resemble an agave. The plant has a distinctive long tuberous root which is characteristic.

Due to its attractiveness to collectors the agave cactus has been reduced in population size, although it has some natural defence from collection through its wide dispersal/sparsity in habitat and hence difficulty for collectors to locate. This also has a negative side as populations, when found, have been cleared by collectors. First described in the mid 1800s it was then lost in cultivation and for a time was a rarity in collections. It grows easily from seed and today there is enough cultivated material to supply the market.

All the CITES recorded trade is in artificially propagated plants. Canada and the United States of America are the main exporters.

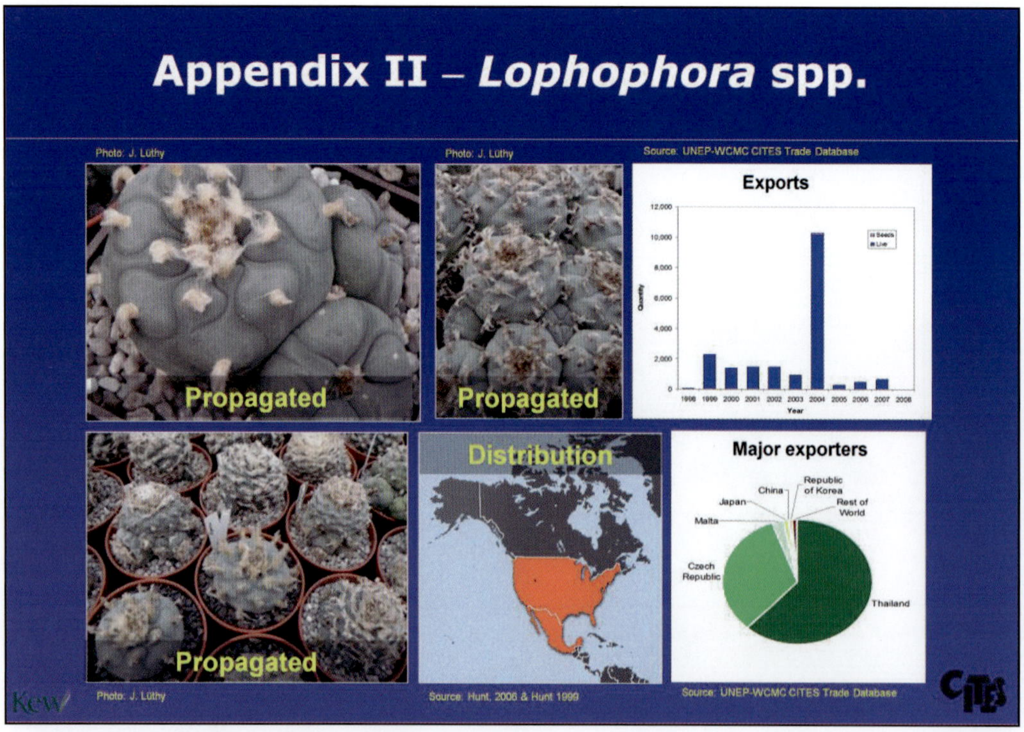

Slide 54: Appendix II – *Lophophora* spp.

The genus *Lophophora* includes three species which occur in Mexico and the USA (Texas).

Lophophora williamsii, whose common name is peyote, is notorious for its hallucinogenic properties. It is a species of easy cultivation and propagation by seed, and illegal CITES trade is unlikely. However, *Lophophora williamsii* is controlled under drug regulations in some countries and possession and trade may be forbidden.

The CITES trade data shows only trade in artificially propagated specimens, mainly in live *Lophophora williamsii*. Thailand and the Czech Republic are the main exporters.

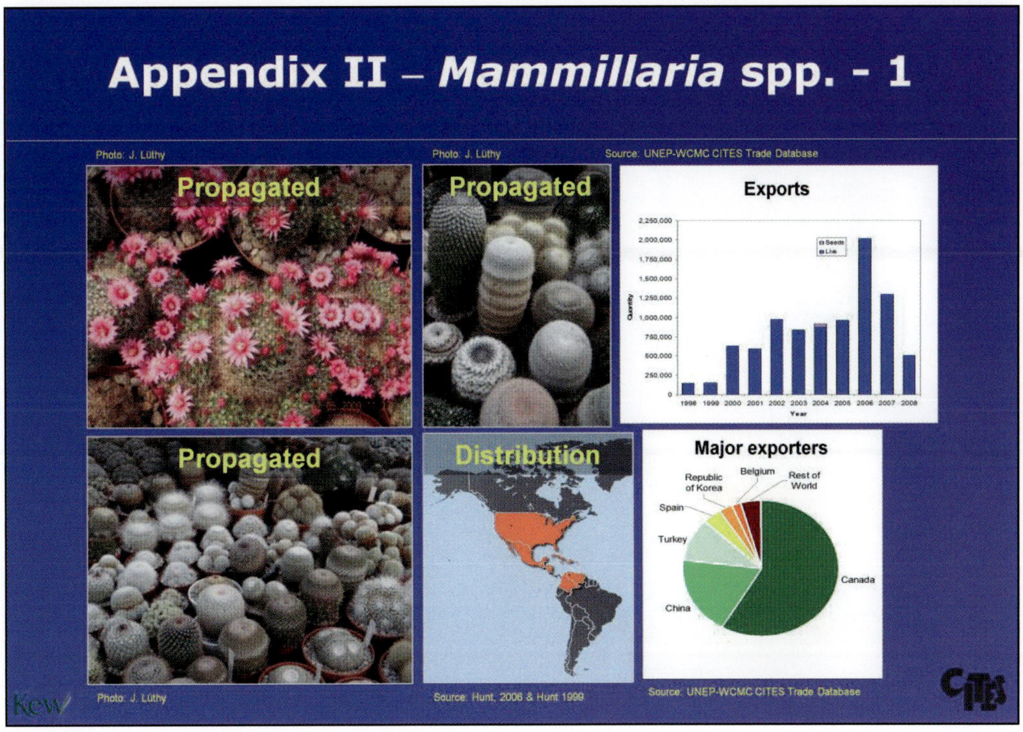

Slide 55: Appendix II – *Mammillaria* spp. - 1

The genus *Mammillaria* contains 163 species and 69 subspecies concentrated in Mexico, but also occurring in south-western USA, the West Indies and South America. Plant shapes may be spherical or short and cylindrical, solitary or freely clustering, with or without milky sap. The small tubercles (wart-like protuberances) range in shape and size but always have an intersecting spiral arrangement and there is often hair between the tubercles. The spines are varied in shape, length and colour.

The great diversity in species and habit make this genus among the most traded cacti. Propagation by seed is easy for most species and there is a large amount of artificially propagated specimens in trade.

Trade is virtually confined to live artificially propagated plants. Significant trade has been recorded in *Mammillaria elongata* and *M. hahniana.* Canada and China are the main exporters.

Slide 56: Appendix II – *Mammillaria* spp. - 2

There are 63 taxa that are considered to be threatened with extinction and one, *Mammillaria columbiana* subsp. *yucatanensis,* is thought to be extinct in the wild. *Mammillaria* is perhaps the most popular genus among cacti enthusiasts and taxa are propagated in their millions across the globe.

Flowers are usually of bell or funnel shape with berry-like fruit most frequently red in colour. The appearance of the plants has given rise to common names such as pincushion cacti.

There has been ongoing low level illegal trade in wild specimens, with some collectors seeking habitat specimens of the rarer or more attractive taxa. Most recently, smuggling is most likely to occur through the mail or courier service as in the case with all of the smaller cacti. Packages from the range States should therefore be targeted for Customs inspection.

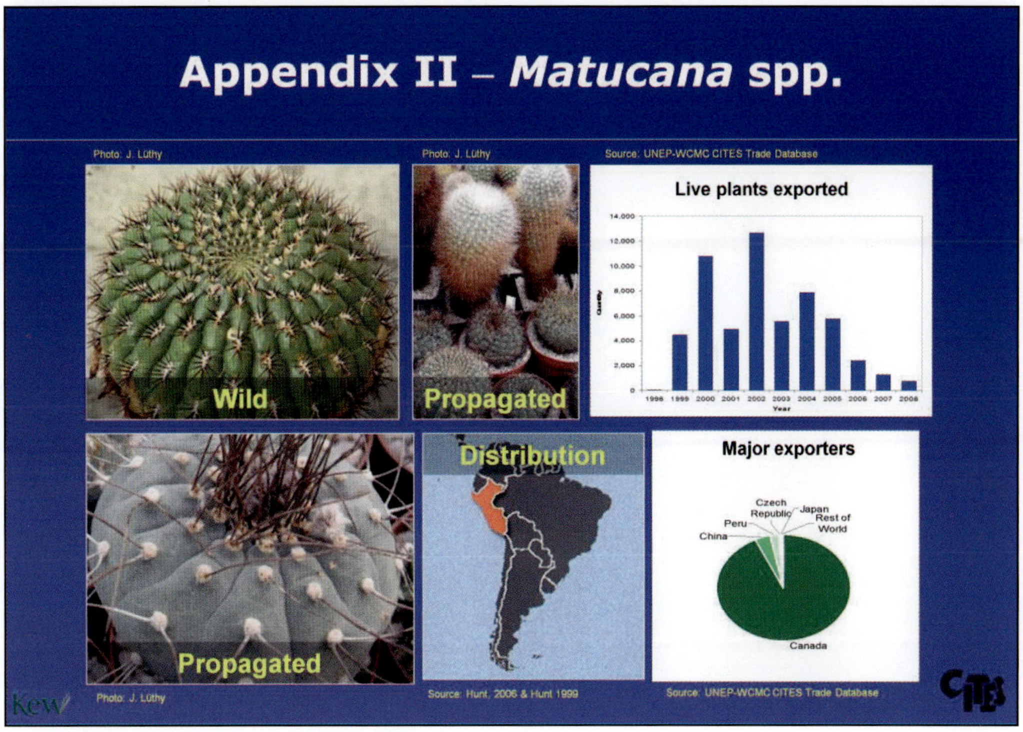

Slide 57: Appendix II – *Matucana* spp.

Matucana is a genus of 14 species and 7 subspecies of globular cacti which is confined to the Peruvian Andes.

The plants are short and spherical or barrel-like. The flowers are often a brilliant red. The striking flowers are day-flowering and there is a large variety within the genus which makes them attractive to collectors. However, cultivation is limited and a number of species have been depleted in habitat due to collection for international trade. Nine taxa are considered to be threatened with extinction.

Artificially propagated material is now available in the market. Some wild-collected plants may be found in trade declared as artificially propagated and care should therefore be taken in inspection of this material.

CITES recorded trade is virtually confined to live artificially propagated plants with only 0.08% declared wild in the period 1998-2008. Main trade is in *Matucana aureiflora* and *M. Madisoniorum*. Canada is the main exporter.

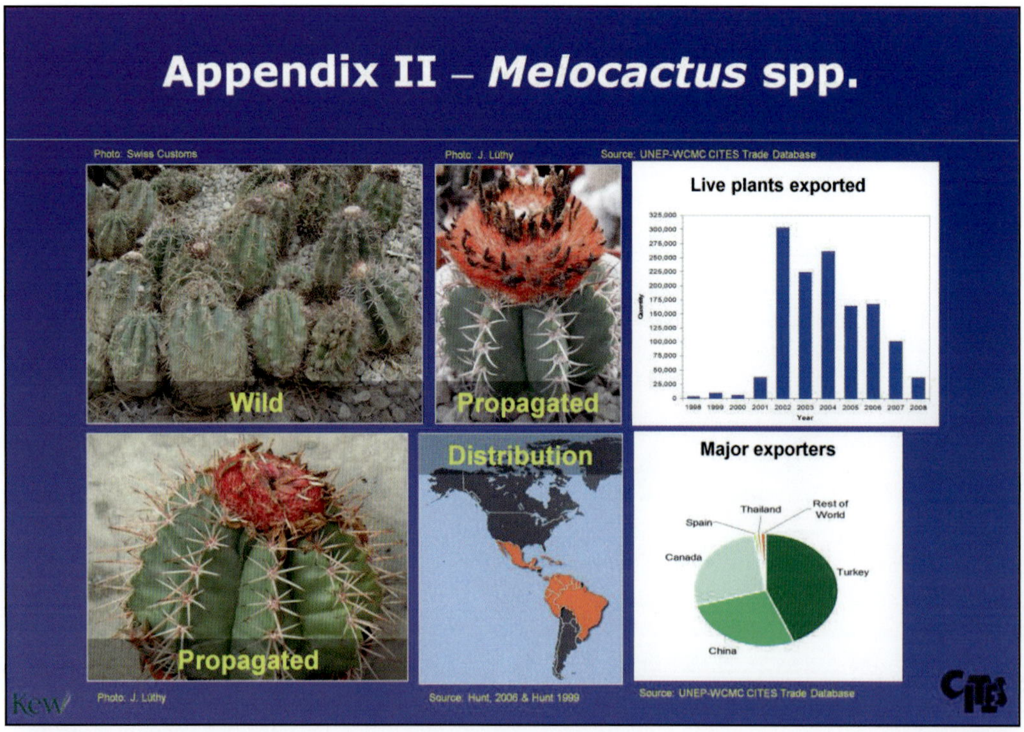

Slide 58: Appendix II – *Melocactus* spp.

The genus *Melocactus* includes 37 species and 13 subspecies distributed in the majority of the islands in the West Indies, southern Mexico, and southwards into South America (especially Brazil). They are globular to short and cylindrical plants and are solitary and occasionally clustering. The height is 0.15-1m and the diameter of the main body is 10-20cm. The body of the plant has 9 to 20 ribs, usually fairly straight, with oval-shaped areoles up to 2.5cm, some of which have spines. The spines, usually stout, are often curved and pressed against the body and can be from 1.25 to 7.5cm long. Mature plants have a terminal cephalium (hat-like flowering zone), which is bristly or woolly.

Seeds are available and there is propagation. The plants are slow in cultivation and propagation is normally by seed which is relatively easy. Slow-growing, they take a long time to reach maturity and produce the characteristic cephalium (hat-like bristly flowering zone) which in some taxa matches or is taller than the body of the plant. Slowness of growth and demand for the mature cephalium-bearing plants have made the most striking species prone to heavy wild collection. Due to this, four species are listed in Appendix I: *Melocactus conoideus, M. deinacanthus, M. glaucescens,* and *M. paucispinus*. All of the Appendix I taxa are endemic to Brazil. Some 20 taxa are considered to be threatened with extinction.

The cephalium makes the genus highly attractive to collectors with significant wild collection in the past. Species are now widely distributed in cultivation and this has substantially reduced the demand for wild plants although some may be found in trade. Wild-collected mature plants usually die after a few years in cultivation as they do not produce new root systems.

CITES trade is recorded as largely artificially propagated material. *M. azureus* and *M. curvispinus* subsp. *caesius* are traded in large volumes. Turkey, China and Canada are the main exporters, and trade has decreased over the years.

Slide 59: Appendix II – *Neolloydia* spp.

The genus *Neolloydia* is composed of just two species of small spherical cacti with poorly developed ribs but well-developed tubercles. It is spread from central Mexico to southern USA (Texas) in desert or semi-desert areas. *Neolloydia matehualensis,* which is only confirmed to occur in San Luis Potosí, Mexico, is considered Vulnerable while the widespread *N. conoidea* is rated as Least Concern.

The boundary of the genus has been poorly defined in the past and some authors have included some 30 taxa in the genus so it has been of some interest to collectors. Trade data may therefore be under-representative of actual trade due to the range of names still used by collectors. Illegal trade in the taxa currently accepted in the genus would appear unlikely.

CITES trade data records all trade as being in artificially propagated live plants. Canada is the main exporter.

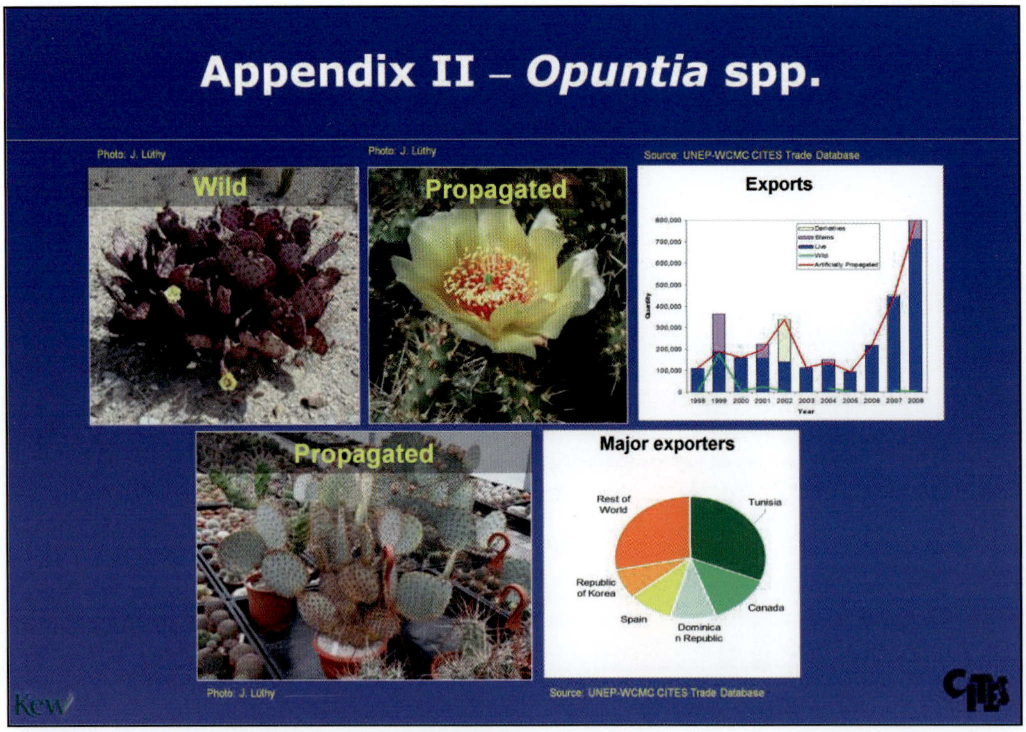

Slide 60: Appendix II – *Opuntia* spp.

The revised genus *Opuntia* is composed of 75 species and occurs in all countries of the range of the Cactaceae family, with the majority of species in the USA and Mexico. Recent work confines the genus to the former subgenus *Platyopuntia* which has species with flatish branch segments. Stems, flowers, and parts and derivatives thereof of naturalised or artificially propagated plants of genera *Opuntia* subgenus *Opuntia* and *Selenicereus* are exempt from CITES.

Among live plants, most trade is in non-regulated *Opuntia ficus-indica* stems. Trade in cultivated or naturalised populations is mostly in *O. microdasys* cultivars. Mature *Opuntia* plants are usually large specimens and it is unlikely that collectors require species of wild origin. Some species such as *O. galapageia* may be of interest to specialised collectors and trade of wild specimens could exist.

CITES recorded trade is in multiple parts and derivatives. This slide illustrates trade in live plants (80% of total trade), stems (12%) and derivatives (7%). There have been high volumes of trade in *Opuntia ficus-indica.* Trade is now virtually confined to live artificially propagated plants but some low levels of wild trade remain. Tunisia is the main exporter.

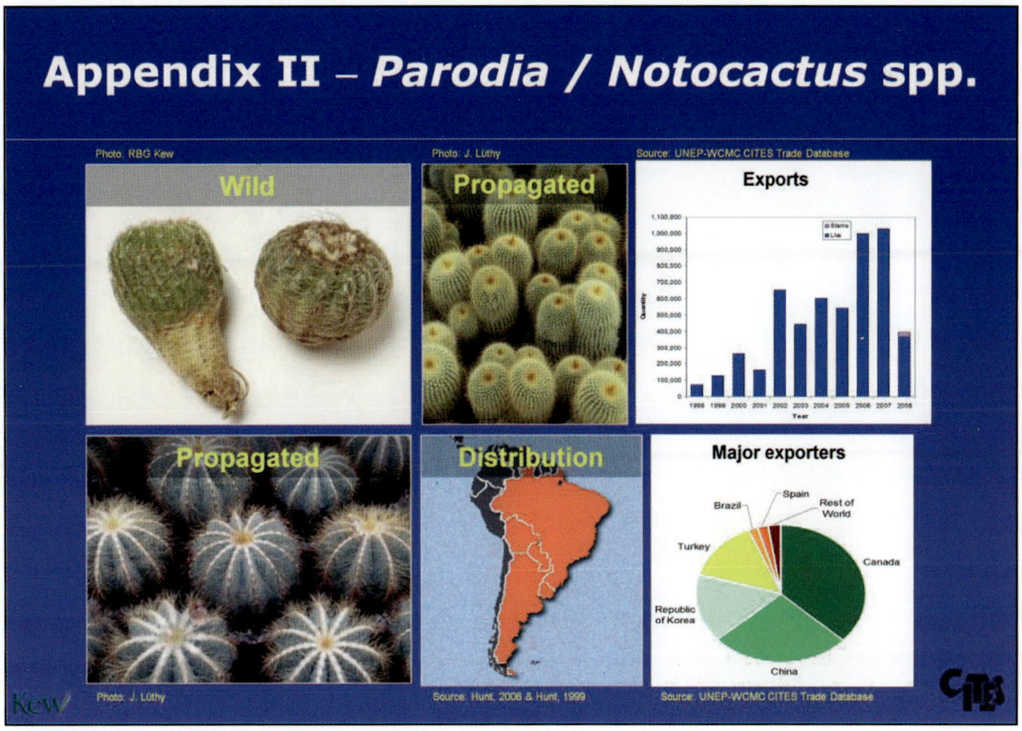

Slide 61: Appendix II – *Parodia (Notocactus)* spp.

The genus *Parodia* comprises 58 species and 8 subspecies and includes the genus *Notocactus*. It is found in Argentina, Bolivia, Brazil, Paraguay and Uruguay. *Parodia* are small, spherical cacti, often ball-like, that flower freely with large brightly coloured blooms of funnel or bell shape. Some 25 taxa are considered to be threatened with extinction. There is no up-to-date treatment of the genus and its boundaries remain disputed with the current tendency to consider it a broad genus with much variation as opposed to previous multiple splits.

Although slow-growing, their small size and brightly coloured flowers, which appear in 2-3 years, have made the group attractive to collectors.

The genus is now widely propagated although there may be some illegal trade of wild-collected plants with collectors searching for the rarer taxa.

Trade data illustrated is for *Parodia* species only. The CITES recorded trade is in artificially propagated live plants, with only a very small amount of wild trade. Canada, China and the Republic of Korea are the main exporters.

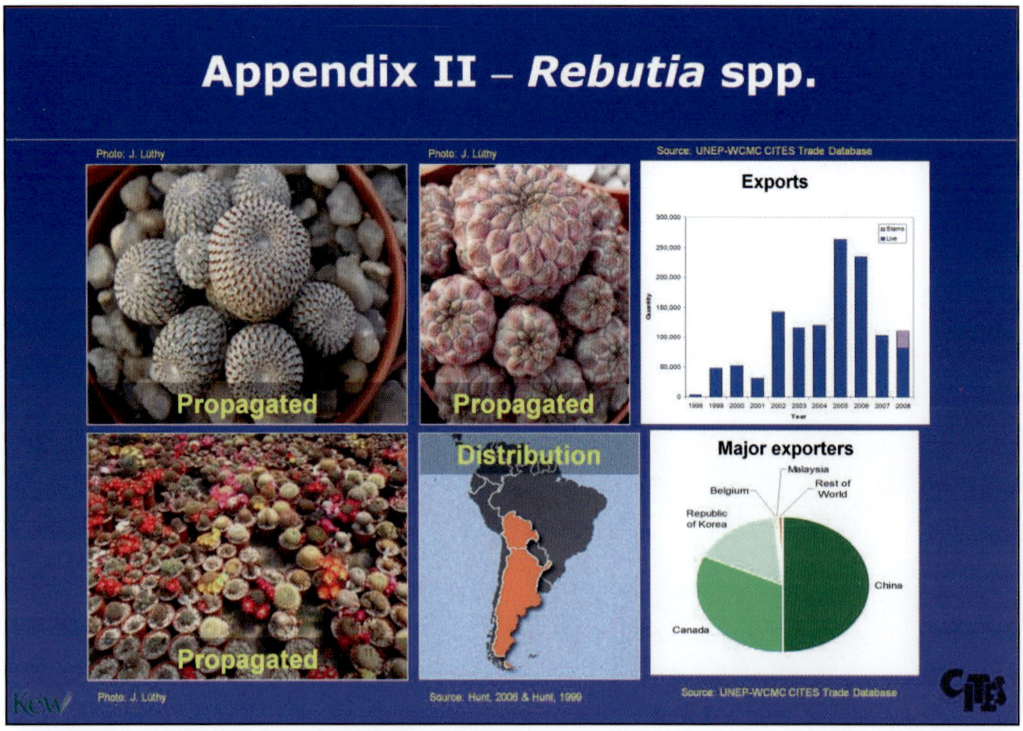

Slide 62: Appendix II – *Rebutia* spp.

The genus *Rebutia* is composed of 29 species and 11 subspecies. It is found in the mountains of Bolivia and Argentina. They have a low-growing habit, are often freely clustering and ribless. They are sparsely spined and free-blooming, with funnel-shaped flowers in a range of colours. Some 18 taxa are considered threatened with extinction.

Rebutia is easy to cultivate and produces a range of brightly coloured flowers and is very popular with enthusiasts. It is now widely propagated in large numbers around the world.

Rebutia comprises five subgroups, *Weingartia*, *Sulcorebutia*, *Mediolobivia*, *Aylostera* and true *Rebutia*, and these names may be used by the trade.

CITES trade recorded to be in 99.99% live artificially propagated plants. *Rebutia marsoneri*, *R. minuscula*, and *R. canigueralii* are the species most frequently recorded in the trade data. China and Canada are the main exporters.

Slide 63: Appendix II – *Rebutia cintia* (*Cintia knizei*)

Rebutia cintia is a sought-after miniature alpine cactus from 4000m in the Andes near Otavi, Bolivia, where it was discovered in 1969. It is a spherical cacti 3-5cm in diameter, spineless, with yellow flowers 3-4cm in diameter born near stem tips and tuberous carrot-like roots.

It quickly entered trade after it was first found and wild plants became available in the international market. CITES trade data (recorded as *Rebutia cintia* and *Cintia knizei)* show a small amount of trade in artificially propagated live plants from the Czech Republic. Cultivated plants, usually grafted, are now available in the market but illegal trade may still occur.

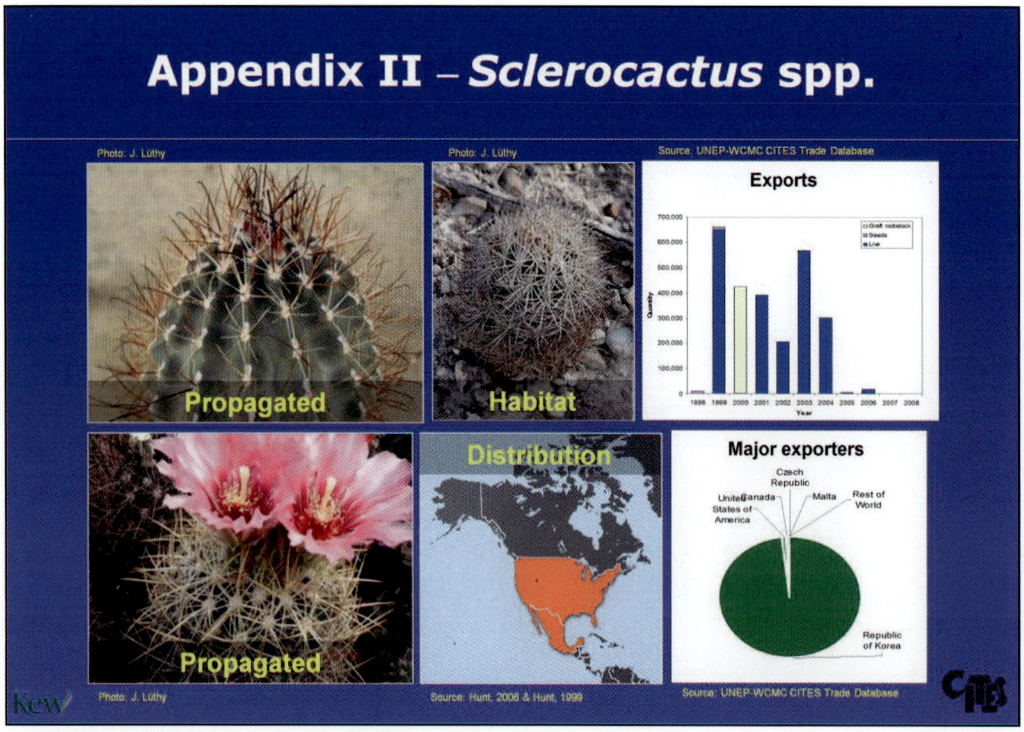

Slide 64: Appendix II – *Sclerocactus* spp.

The 20 species of the genus *Sclerocactus* originate from the south-western USA and northern Mexico. The cylindric plants are globular to short, up to 45cm tall and 10cm wide. Some species however are dwarf, and barely emerge from the ground so are difficult to spot. The plant body ranges in colour from green to bluish-green. Most species have an interlacing spine structure.

Eight taxa are included in CITES Appendix I since 1983 and one since 2003.

CITES trade data is virtually confined to seeds and artificially propagated live plants, with a small amount of wild exports in 1998, 2005, 2008 and 2009 and a significant amount of root stock exported in 2000 from the Republic of Korea. *Sclerocactus uncinatus* subsp. *uncinatus* is most frequently seen in trade. The USA and the Republic of Korea are the main exporters.

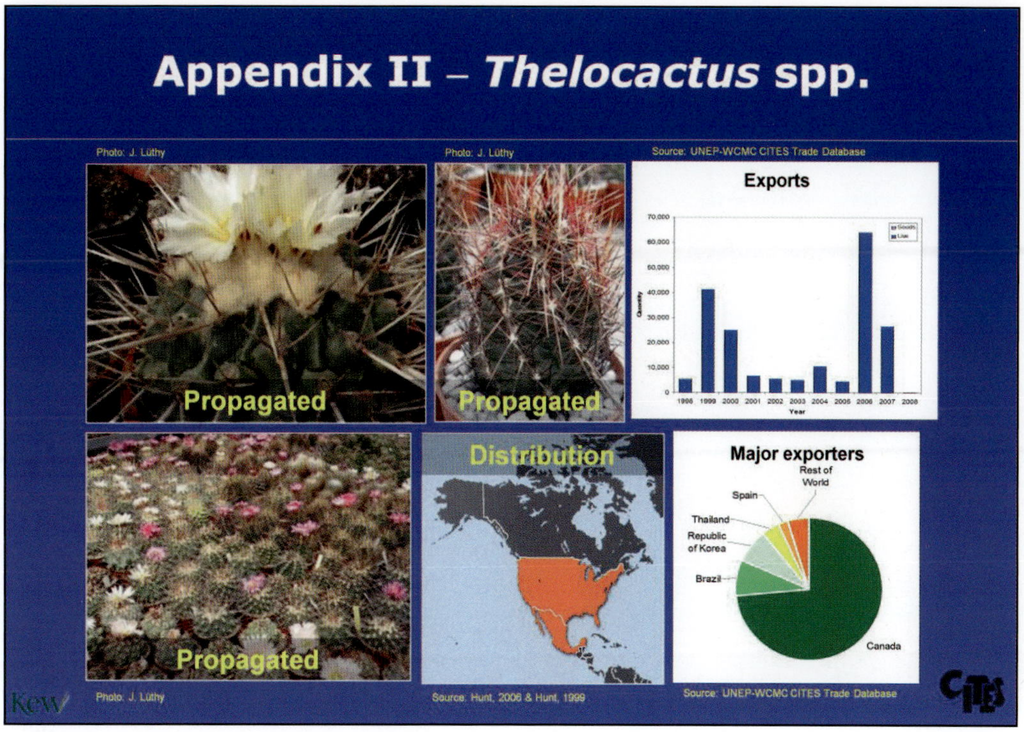

Slide 65: Appendix II – *Thelocactus* spp.

The genus *Thelocactus* contains 14 species and 6 subspecies located in Mexico and the USA (Texas). Nine taxa are considered to be threatened with extinction. The genus has stems in single or clustered habit, and is spherical or barrel-like with distinct tubercles. The spines are variable and the stem may be ribbed, often in a spiral fashion. The plants bear attractive numerous flowers and are relatively easy to cultivate, usually from seed. These characteristics make it very attractive to collectors.

The genus is now widely available as cultivated stock and illegal trade is unlikely.

CITES trade data is virtually confined to artificially propagated live plants, with *Thelocactus setispinus* frequently seen in trade. Canada is the main exporter.

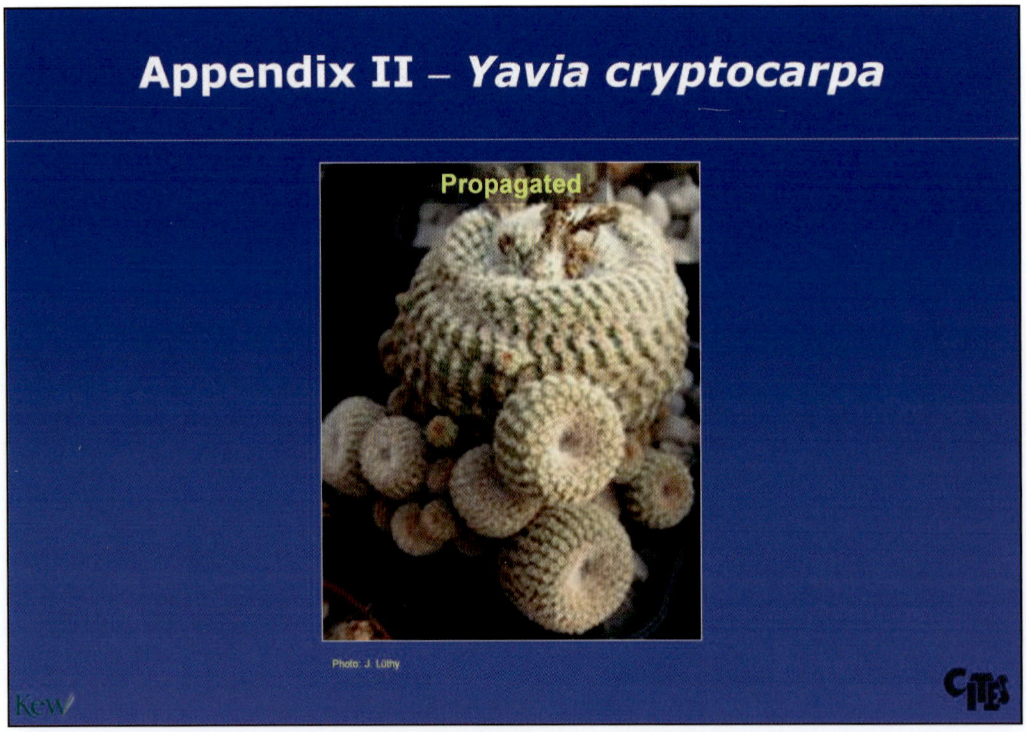

Slide 66: Appendix II – *Yavia cryptocarpa*

The monotypic genus *Yavia* is confined to semi-desert at 3700m in Yavia, Quebrada de Torqueros, Argentina, close to the border with Bolivia. It was discovered in late 2001. The plant is nearly completely subterranean with flattened stems up to 3cm in diameter, with the head of stems frequently very small. The stems swell and break the surface following rains. Spines are small and arranged like the teeth of a comb. Flowers are white to pink.

The species is considered to be Critically Endangered. The plant is propagated by seed and frequently grafted in an attempt to speed growth.

Little or no trade is recorded in the CITES Trade Database, apart from a small amount of artificially propagated live plants exported from Malta, and the origin of any plant found in trade should therefore be checked.

Slide 67: Epiphytic Cacti

An epiphyte is a non-parasitic plant that grows on another plant or surface. Lacking an extensive root system, it gains its water and nutrients from the air or from water streaming over its surface. Epiphytic cacti grow in the neotropical forests and woodlands. Some genera of epiphytic cacti do not root in the soils and grow only on trees branches, e.g. *Disocactus, Epiphyllum* and *Rhipsalis,* while other are facultative epiphytes and may root and start growing in the soil, later becoming epiphytic, like *Hylocereus* and *Selenicereus.* Epiphytic cacti form adventitious roots very easily and these anchor the plants to the branches in addition to absorbing minerals and water.

Stems of epiphytic cacti are quite different from those of terrestrial cacti. They are broad and leaf-like, and ribs are thinner to the point that they cannot support an upright growth without anchoring themselves to trees. Facultative epiphytes possess broader stems, cylindrical to triangular which may support some upright growth before anchoring to trees.

Flowers differ greatly in epiphytic cacti. Some species are night-flowering, and in this case flowers are white and large, as in 'Queen of the Night' (*Selenicereus grandiflorus*) whose dramatic flowers may grow up to 40cm in diameter and are pollinated by bats. Day-flowering species usually have smaller flowers ranging in colour from white to bright red.

In cultivation epiphytic cacti require some shade and higher humidity than other cacti. Propagation by cutting and grafting is very easy and trade in wild-collected plants is very unlikely. Only a few specialist collectors grow epiphytic cacti. Most of the trade is in those hybrids and cultivars which are propagated in huge quantities and for this reason are excluded from CITES regulation. The most widely known and grown epiphytic cacti are Thanksgiving and Christmas cacti (both hybrids of *Schlumbergera*), and Easter cacti (*Rhipsalis gaertneri* or *Hatiora gaertneri* – formerly *Rhipsalidopsis gaertneri*). These are high volume traded supermarket cacti.

Slide 68: Columnar Cacti

Within the Cactaceae family there are several columnar species that may be very tall, up to 10m and over. These species are not very common in trade as mature specimens. Some specimens of the saguaro (*Carnegiea gigantea*) may be found in trade as wild plants with tags stating that they were plants collected due to road construction or land clearance.

Most columnar species are traded as young plants propagated from seed. Among these *Espostoa lanata* (far right) and *Oreocereus celsianus* (far left) are common because of the very hairy stems which results in the common name of 'old man cacti'.

Non-CITES Cacti
Kew

Slide 70: Leaf-bearing Cacti

Some cacti have well developed leaves, and they belong to the genera *Pereskia*, *Pereskiopsis* and *Quiabentia*. Since there is little trade in the leaf-bearing cacti these three genera have been removed from CITES regulations. The genus *Pereskia* is the best known example of this group of cacti.

Slide 71: Exempted Hybrids and Cultivars

CITES excludes cacti colour mutants from regulation. Colour mutants cannot survive after germination in habitat and therefore can only be artificially propagated via grafts.

Schlumbergera is an epiphytic genus very common in cultivation. Several specialised growers are able to select cultivars that produce very attractive flowers. *Opuntia microdasys* is also cultivated by this method, but areole colour can also be selected. *Schlumbergera* and *Opuntia microdasys* cultivars have lost any relation to wild populations, both are very easily propagated by cuttings, and are therefore excluded from CITES controls.

Implementing CITES for Cacti

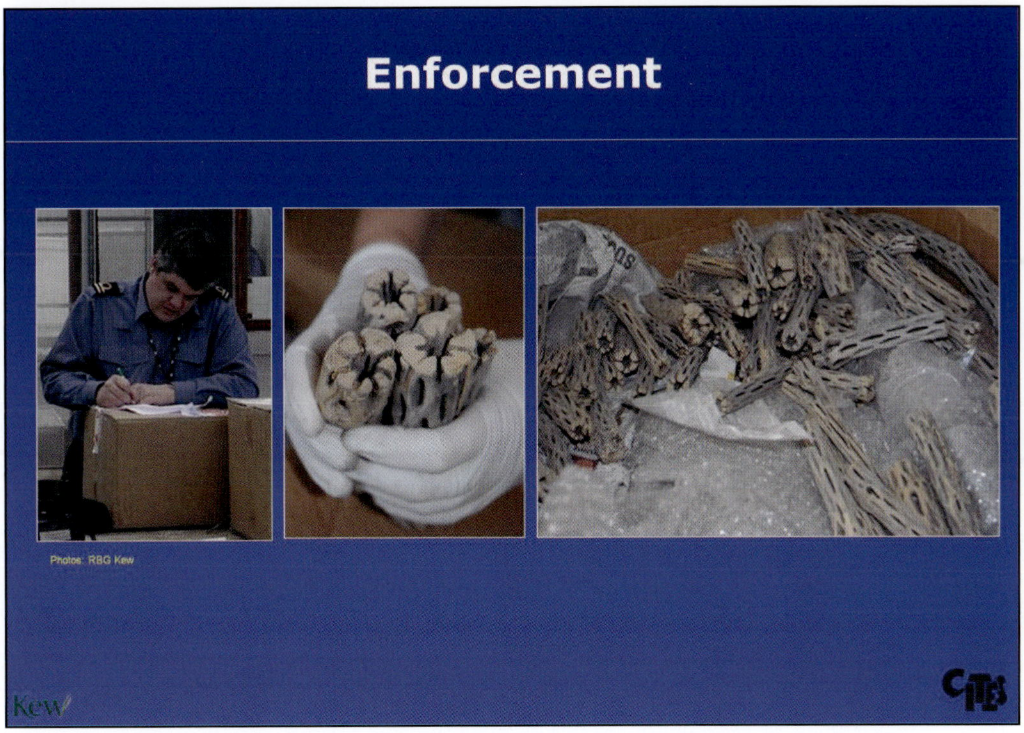

Slide 73: Enforcement

The enforcement of CITES controls is carried out at different levels. Within an exporting country it is carried out by the inspection of nurseries, traders, markets and, less frequently but most importantly, of the plants at time of export. Inspections can also occur at the time of import and post-importation in the major trading countries. Enforcement agencies also survey trade shows, advertisements in the trade press and the internet.

A small number of countries have enforcement teams specially trained to identify CITES specimens – animals or plants. CITES enforcement for plants is most likely to be carried out by general Customs staff or by officials trained in plant health controls. When CITES enforcement is carried out by general Customs staff the enforcement procedures are concentrated on the documentation, not the plants. Thus, Customs may check to see if the permits are correctly filled in, stamped and issued by the correct authorities. They also check other documents and invoices to see if any CITES material named on the accompanying documentation is missing from the CITES permits.

Where such general Customs staff are used to check CITES plants it is vital that they have contact with a centre of expertise on the identification and conservation of plants. Such a centre should be the national Scientific Authority. However, in some cases the national Scientific Authority may be a committee or a government department with expertise centred on animals. In this case, the enforcement authorities should build a relationship with a national or local botanic garden or herbarium. Such a relationship is vital.

The Customs Officers will need some basic training on the plants and parts and derivatives covered by CITES and will need help on targeting detrimental trade. Most importantly, Customs officials will need access to experts who can identify CITES plants. Such experts can also advise on, and have access to, facilities for holding seized or confiscated material. These scientists may be called on to be expert witnesses, which are vital if breaches of the controls result in prosecution and court appearances.

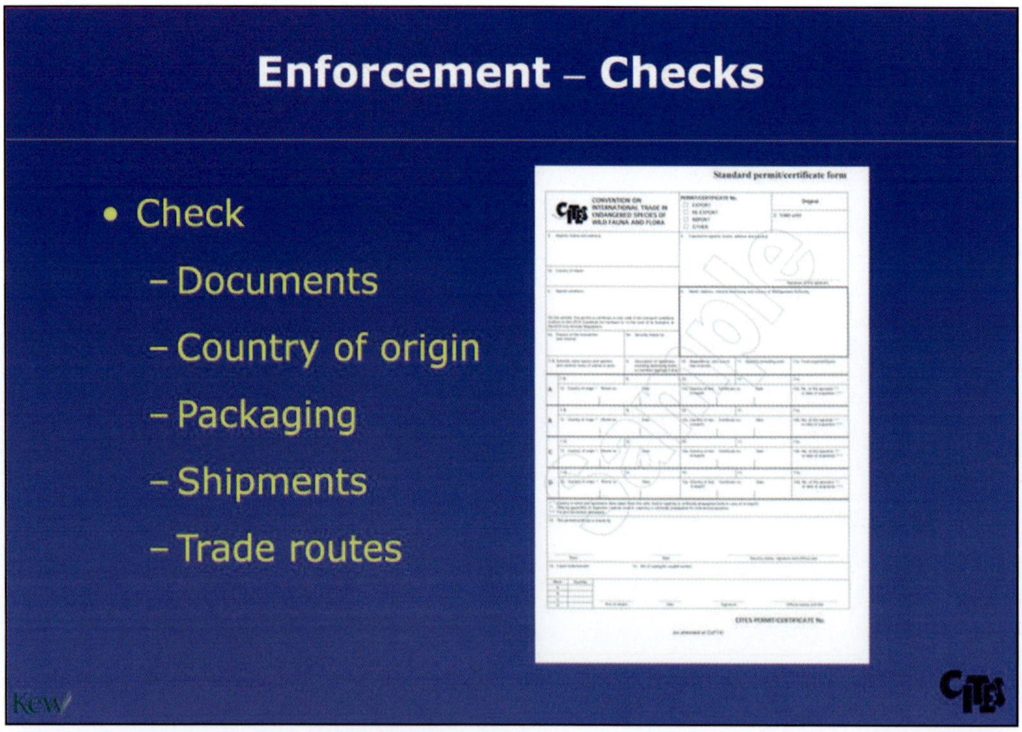

Slide 74: Enforcement - Checks

Documents – Check the authenticity of the CITES permits (signatures, stamps), and check the plant names and number of specimens on the permit against the delivery note or invoice. Also, check the source of the plants – are they declared as wild or artificially propagated?

Country of origin – Always check the country of origin on the permits. Are the succulents being exported from a country where the plants grow in the wild? If so then the plants may be more likely to be wild-collected. Most countries have banned the export of wild-collected plants. Countries may express concern over the illegal export of their wild-collected succulent plants and ask for the assistance of other CITES Parties and non-Parties to control this trade. Normally, such a request is published as a Notification to the CITES Parties (you can find this on the CITES website: www.cites.org).

Packaging – Nurseries will usually wrap and package their plants carefully to avoid damaging them. They are then shipped in boxes marked with the nursery's name and with printed labels. Consignments of illegally collected plants may be poorly wrapped using local materials, contain handwritten labels (sometimes with collecting data), and the plants may not be identified to species level to disguise the fact that new unnamed species may have been collected.

Consignments of plants – Collections of illegal plants usually consist of small samples of plants of different size and age groups that are not uniform in shape. They may be damaged (broken or snapped roots), and soil and weeds or native plants may be present amongst the stems and roots. Artificially propagated plants will be uniform in size and shape and be clean of soil, pests and diseases, weeds or native plants.

Trade routes & smuggling – Illegal collections of rare or new species may be shipped using the postal / courier services or hand luggage to avoid detection. Collections may also be split up and sent in several different packages to ensure both a high level of survival and that at least some of the plants will evade discovery.

Slide 75: Wild-Collected – Cacti Smuggling

Tourists visiting countries where cacti are found may return with plants collected in the wild. Some may have collected specimens as souvenirs unaware of CITES restrictions. Others may be semi-professional collectors who collect specimens for commercial reasons; a suitcase filled with wild-collected Mexican cacti can have a commercial value higher than the cost of travel. Luggage of those returning from some countries rich in high-value cacti such as Mexico, Peru, Chile and Brazil, amongst others, should be checked to discourage this form of illegal trade. These slides display wild-collected cacti detected by alert Customs officers in travellers luggage returning to Europe and routing through Poland.

Slide 76: Wild-Collected – Cacti Smuggling

Another method of smuggling wild-collected cacti specimens is via the postal or courier services. Some of the high-value cacti are small to medium size plants and can easily be hidden in parcels which contain other items such as books, DVDs or video cassettes. Cacti have even been found smuggled in cigarette packages (see slide). Cacti are sometimes stuffed into cardboard tubes used for packaging maps and prints. Postal and courier parcels originating from countries rich in high-value cacti such as Mexico, Peru, Chile and Brazil among others, should be checked to detect and discourage this form of illegal trade.

Wild or Artificially Propagated – Key Characteristics

	Wild	Artificially propagated
General Appearance	• Irregular shape & size • Wounds/ insect damage	• Uniform • Healthy plant parts
Spines	• Irregular & broken • Thicker	• Uniform & intact • Thinner & weaker
Roots	• Irregular • Dead & broken • Cut back	• In shape of pot • Roots cut back but healthy
Soil	• Native soils and associated plants	• Usually clean of soil • Horticultural soil present (e.g. peat, sand, perlite, rockwool)

Slide 77: Wild or Artificially Propagated – Key Characteristics

Distinguishing between wild-collected and artificially propagated plants is not a straightforward matter, but there are certain characteristics that can be used to make this distinction. This slide highlights some of those characteristics.

Wild-collected plants carry the marks of growing in their natural habitat. Plants propagated in nurseries bear the marks of an artificial, well maintained environment. They are clean, uniform and packed to a high standard. Propagated plants are uniform in size, while those wild-collected may be a mixture of small to large specimens.

Plant appearance is also very important. Plants in the wild are exposed to the elements such as rain and wind and are subject to attack from herbivores, which may damage stems and spines. Plants propagated in a nursery are in a controlled environment without such disturbances and plants are healthy with intact spines and clean stems.

When taking plants in the wild, collectors do not pay attention to the root system; they simply want to collect a large number of plants to sell. They often pull the plants out of the soil without digging, and as a result wild-collected plants usually have damaged roots or may not have roots at all. Horticulturalists and growers pay a lot of attention to the root system as it is essential for a fast-growing and healthy plant. Artificially propagated plants have intact roots, or when they are cut back they are healthy.

Commercial shipments of artificially propagated plants are usually clean of soil or horticultural soil may be present. Wild-collected plants usually have some native soil attached to them, which is usually quite distinct from soil used in cultivation.

It is important that you call in an expert to check the status of any plants you consider may be wild-collected rather than artificially propagated.

Slide 78: Wild or Artificially Propagated?

Some nurseries may try to 'clean' wild-collected plants, cultivating them for some time before selling. A new root system may develop again and some new growth may appear. However, the root system of such plants will be different to those of artificially propagated plants: the site of the original cut will be evident and the new roots will be thinner. Also, the colour of new growth will be different to that of growth in the wild and spines will be thinner and weaker.

This slide illustrates some of the characteristics that may be displayed by wild-collected and artificially propagated plants.

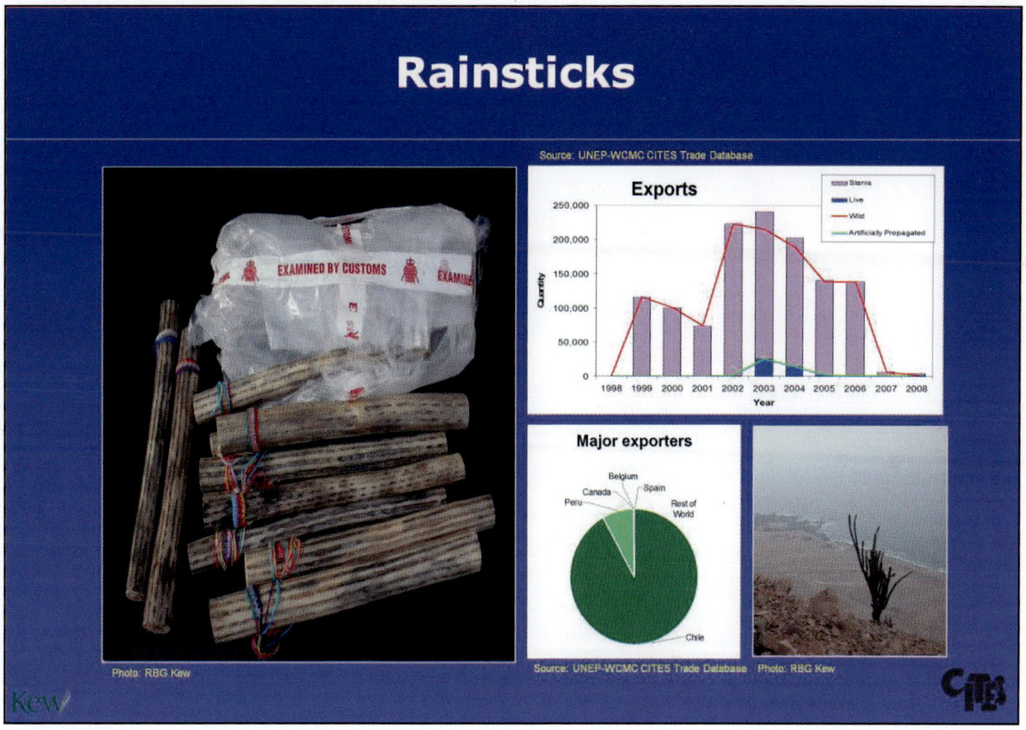

Slide 79: Rainsticks

Rainsticks are novelty items that have become popular as tourist souvenirs for visitors to Latin America. However they are now found in craft shops across the world, and frequent seizure of specimens by customs officers at airports has resulted in a special CITES 'personal effects' exemption for a small number of specimens per person.

Rainsticks are made from the dried stems of columnar cacti, most frequently *Echinopsis chiloensis* and *Eulychnia acida.* Dry dead stems are harvested in the field, cut and cleaned. If still present the spines may be pushed inward or artificial baffles inserted, along with coarse sand or small stream-washed pebbles. The ends are sealed, the stems waxed or varnished and decorated with coloured ties – frequently in national colours. When shaken the rainsticks make a soft 'waves on the seashore' sound.

International trade expanded with the growing interest in traditional crafts and there was some concern that the harvest of specimens might go beyond the collection of dead material in habitat to the destruction of large mature cacti. However, the CITES Plants Committee reviewed the trade and found no evidence of detriment. The slide displays trade in stems, all wild-sourced from 1998 to 2008. Exports peaked in 2003 at 222,000 stems and have since declined to under 2,000 in 2008. Chile and Peru are the main exporters and the United States of America, France, United Kingdom and Italy the main importers.

Recently there has been some concern expressed by experts that there may be some element of detriment in the trade, but evidence has yet to be put to the CITES Plants Committee. The Plants Committee reviews such trade on a regular basis as part of the Review of Significant Trade in Appendix II plants and animals.

CITES recorded trade is nearly all in stems rather than live plants. The species included in the trade data are: *Armatocereus matucanensis, A. procerus, Browningia candelaris, Corryocactus brevistylus, Echinopsis atacamensis, E. atacamensis* subsp. *pasacana, E. chiloensis, E. cuzcoensis, E. pachanoi, E. peruviana, Eulychnia acida, Neoraimondia arequipensis* and *Weberbauerocereus rauhii.* Trade data includes exports of stems and live plants. Peaking in 2003, exports in wild specimens, and in the species overall, have declined. Chile is a major exporter, followed by Peru.

Slide 80: Seeds

Seeds of Appendix I Cactaceae spp. are regulated by CITES, while those of Appendix II are not. However, seeds of Mexican cacti are subject to CITES regulation and require an export permit.

Slide 81: Fruits

Some 80 species of cacti have fruit which are commercially important nationally or internationally. Prickly pear (*Opuntia ficus-indica*) is far the best known but some columnar and epiphytic cacti are becoming more popular worldwide.

Opuntia ficus-indica fruits are produced both in their native country (Mexico) and in several subtropical and Mediterranean countries, where they are cultivated and in some instance naturalised. Mexico produces over 345,000 tonnes of fresh fruit per year, and Italy some 70,000 tonnes.

Another group of edible fruit-producing cacti is that of the facultative epiphytic *Hylocereus*. They produce medium to large fruits with coloured peel and white or coloured, juicy pulp and are known as dragon fruit. They contain small digestible seeds and lack the fine, barbed bristles that make prickly pear fruits unpleasant to handle. Native to tropical regions of North and South America, they are now grown commercially in Nicaragua, Columbia, Vietnam, Mexico and Israel. Recent areas of production are Malaysia, Taiwan, Australia and USA.

Some columnar cacti are also cultivated for fruit production, mainly the genus *Stenocereus* and their fruits are usually referred to as pitaya. Cultivation is presently limited to Mexico, near their natural habitats. *Stenocereus queretaroensis* and *S. stellatus* are the most commercially cultivated species, with fruits up to 200g, having a very tasty white to purple pulp, rich in Vitamin C. Trade in this genus is usually very local.

In recent years the commercial production of cacti fruit is increasing, and new species, such as *Cereus*, are being tested for commercial exploitation. All fruits produced from artificially propagated or naturalised cacti are excluded from CITES regulation.

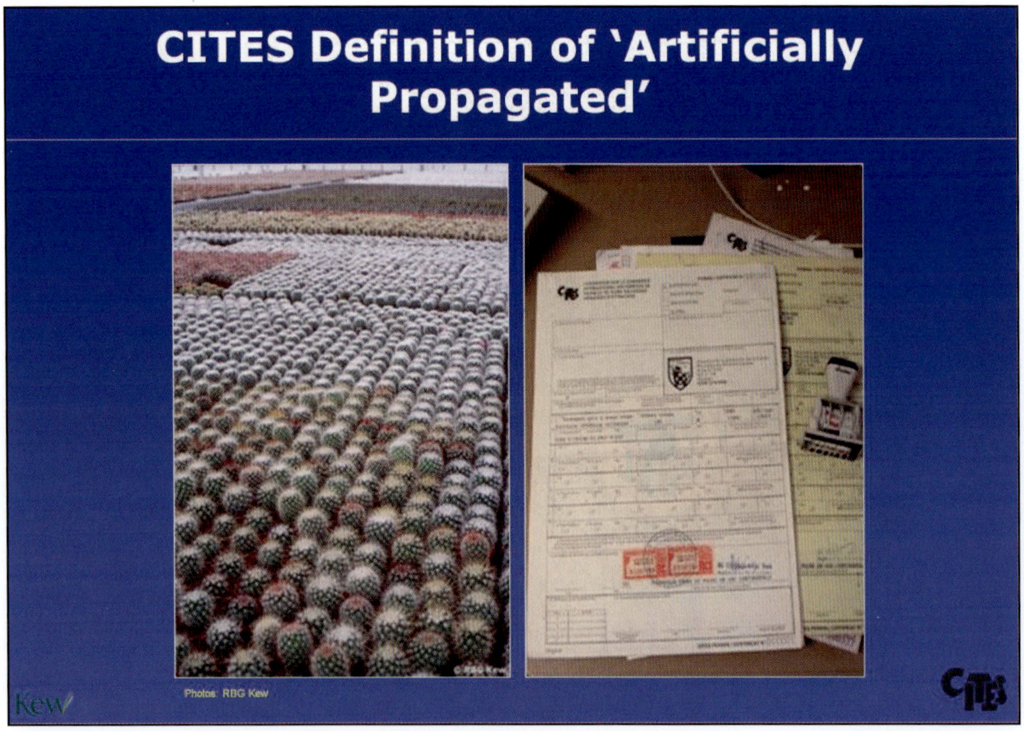

Slide 82: CITES Definition of 'Artificially Propagated'

The CITES definition of artificially propagated is included in Resolution Conf. 11.11 (revised CoP15 – check cites.org for details) *Regulation of trade in plants*. The definition within CITES includes several unique criteria. The application of these criteria may result in a plant which bears all the physical characteristics of artificial propagation being considered as wild-collected in CITES terms. Some key points are:

• *Plants must be grown in controlled conditions*. This means, for example, the plants are manipulated in a *non-natural environment* to promote prime growing conditions and to exclude predators. A traditional nursery or simple greenhouse are both 'controlled conditions'. A managed tropical shade house would also be an example of 'controlled conditions'. Temporary annexation of a piece of natural vegetation where wild specimens of the plants already occur would not be 'controlled conditions'. Also, wild-collected plants are considered wild even if they have been cultivated in controlled conditions for some time.

•The cultivated parent stock must have been *established in a manner not detrimental to the survival of the species in the wild* and managed in a manner which *ensures long term maintenance of the cultivated stock.*

•The cultivated parental stock must have been *established in accordance with the provisions of CITES and relevant national laws.* This means that the stock must be obtained legally in CITES terms and also in terms of any national laws in the country of origin. For example, a plant may have been illegally collected within a country of origin then cultivated in a local nursery and its progeny exported declared as artificially propagated. However, such progeny cannot be considered to be artificially propagated in CITES terms due to the illegal collection of the parent plants.

•Seeds can only be considered artificially propagated if they are taken from plants which themselves fulfil the CITES definition of artificially propagated. The term *cultivated parental stock* is used in order to allow some addition of fresh wild-collected plants to the parental stock. It is acknowledged that parental stock may need to be occasionally supplemented from the wild. As long as this is done in a legal and sustainable fashion it is allowed.

Applying the CITES definition is a complex mixture of checking legal origin, propagation status and non-detrimental collection. To achieve this, the assessment needs to be carried out in close co-operation between the CITES Management and Scientific Authorities. The implementation of the criteria on a day by day basis needs to be tailored to the situation in an individual CITES Party. National CITES authorities should consider producing a checklist as a means of standardising the process and informing the local plant traders.

Slide 83: Cacti in Medicine and Religion

Cacti form part of traditional medicine and spiritual use for a wide range of communities and for western medicine are a source of chemical substances used in medicine and pharmacology (such as alkaloids, lecithins and glycosides). *Lophophora* (peyote) has a history of religious and medicinal use by Native Americans and is also a source of experimentation due to the presence of mescaline. Its use is legal for certain native Indian tribes for cultural purposes in the USA.

A range of other cacti are used in Mexico for healing including species of *Mammillaria*, *Ariocarpus*, *Coryphantha* and *Echinocereus*. The San Pedro cactus (*Echinopsis pachanoi*) has been used in the Andes for healing for thousands of years. More recently there is a surge in interest in plants that provide 'legal highs' and peyote and the San Pedro cactus are sought after as sources of mescaline with numerous web sites offering advice and sale. In many countries it is not illegal to possess the cacti but it is illegal to extract the alkaloid. This interest is stimulating illegal trade in these taxa and the trade requires close monitoring to detect whether it is for horticultural use or illegal use as a drug.

'Nopal Cactus' is a powdered extract of species of Opuntia and is marketed globally as a food supplement. It is claimed to be an appetite suppressant, lower blood sugar, reduce cholesterol and remove and prevent blood vessel plaque. Medical trials have been carried out to test its potential use in mainstream medicine. It is in trade in significant amounts. Cacti extracts are also used as remedies in homeopathic medicine including *Selenicereus grandiflorus* as a treatment for angina.

Slide 84: Contacts & Further Resources

There is a wide range of material available on the classification, conservation, cultivation, distribution, use and, to a lesser degree, trade in cacti. This material is aimed at all interest groups across the amateur and professional spectrum. For those interested in the trade and its regulation a good entry point is the CITES website (www.cites.org). This contains information on standard references used by CITES for the names of cacti, identification guides, papers submitted to the CITES Plants Committee and a record of the committee's discussions. TRAFFIC, the non-governmental wildlife trade monitoring network (www.traffic.org), reports on trade in plants and animals and is a good source of information on recent seizures by enforcement agencies around the world.

There are a wide range of websites dedicated to amateur growers, while professional research institutions also provide a range of very valuable web-based resources. All of these are too numerous to go into in detail here and it is easier to explore the web. However, solid scientific work is needed to check on key data. By far the most useful of the books published in the last ten years is The New Cactus Lexicon, edited by David Hunt (DH Books 2006), which is two volumes, the first containing comprehensive descriptions of the cactus family and the second with complete photographs of the species. It is a masterwork and a vital reference.

For a good general reference Ted Anderson's The Cactus Family (Edward F. Anderson, 2001, Timber Press, Portland, Oregon) remains a good buy.